Creating Sounds from Scratch

Creating Sounds from Scratch

A PRACTICAL GUIDE TO MUSIC SYNTHESIS FOR PRODUCERS AND COMPOSERS

Andrea Pejrolo

Scott B. Metcalfe

UNIVERSITY PRESS

Oxford University Press is a department of the University of Oxford. It furthers
the University's objective of excellence in research, scholarship, and education
by publishing worldwide. Oxford is a registered trade mark of Oxford University
Press in the UK and certain other countries.

Published in the United States of America by Oxford University Press
198 Madison Avenue, New York, NY 10016, United States of America.

© Oxford University Press 2017

All rights reserved. No part of this publication may be reproduced, stored in
a retrieval system, or transmitted, in any form or by any means, without the
prior permission in writing of Oxford University Press, or as expressly permitted
by law, by license, or under terms agreed with the appropriate reproduction
rights organization. Inquiries concerning reproduction outside the scope of the
above should be sent to the Rights Department, Oxford University Press, at the
address above.

You must not circulate this work in any other form
and you must impose this same condition on any acquirer.

Library of Congress Cataloging-in-Publication Data
Names: Pejrolo, Andrea. | Metcalfe, Scott B.
Title: Creating sounds from scratch : a practical guide to music synthesis for producers and composers /
Andrea Pejrolo, Scott B. Metcalfe.
Description: New York, NY : Oxford University Press, [2017] | Includes index.
Identifiers: LCCN 2016019139 | ISBN 9780199921874 (cloth : alk. paper) |
ISBN 9780199921898 (pbk. : alk. paper) | ISBN 9780199921904 (companion website)
Subjects: LCSH: Synthesizer (Musical instrument)—Instruction and study. | Software synthesizers.
Classification: LCC MT724 .P47 2017 | DDC 786.7/13—dc23
LC record available at https://lccn.loc.gov/2016019139

To my dearest wife Irache and beautiful daughter Alessandra.
—Andrea Pejrolo

To my sons, Nathaniel and Andrew, and my beautiful wife Sara.
—Scott B. Metcalfe

CONTENTS

Acknowledgments • xv
Introduction • xvii
About the Companion Website • xix

1 How Did We Get Here? • 1
  Introduction • 1
  Electronic Music in the Twentieth Century • 3
    Musique Concrète • 3
    Tape-Based Composition • 5
    Electronic Music Synthesis • 6
  Instruments • 6
    Theremin • 6
    Hammond Organ • 8
    RCA Electronic Music Synthesizer • 10
  The Rise of the Modern Synthesizer • 11
    Moog • 11
    Buchla • 13
    Mellotron • 14
  The Road to Polyphony • 15
    Duophonic/Duophony • 16
    Paraphony/Paraphonic • 17
    Arrival of True Polyphony • 18
  Musical Instrument Digital Interface • 19
  Digitally Controlled Oscillator • 21
    Synclavier • 21
    Chowning and the FM Synthesizer • 22
  Digital Sampling • 23
  Wavetable Synthesis • 25
  Korg M1 Workstation • 27
  Conclusion • 28
  Acknowledgments • 29
  Bibliography • 29

2 Understanding Sound and Hearing • 31
  Introduction • 31
  What Is Sound? • 31
    Timbre—Harmonic Structure • 32
    Harmonics, Partials, and Overtones • 34
    Natural Harmonic Series in Music • 36

Complex Waveforms • 36
Phase • 36
Pitch • 38
Doppler Shift • 41
Amplitude • 42
Timbral Shape Over Time • 43
Human Hearing • 43
Phantom Image • 43
Critical Bands • 44
Frequency Masking • 45
Loudness • 46
Loudness Relative to Harmonic Content • 46
Loudness vs. Duration • 48
Loudness Relative to Frequency • 48
Loudness Perception (RMS/Peak) • 49
Missing Fundamental • 50
Conclusion • 51

3 The Tools of the Trade • 53

Introduction • 53
Sound Generators • 53
Oscillator (VCO) • 53
Sine Wave • 55
Sawtooth Wave • 56
Noise • 57
Digital Playback • 59
Wavetable Synthesis • 59
Sample Replay/Sampling • 60
Sound Modifiers • 60
Frequency-Based Modifiers • 61
Low-Pass Filter • 61
Sound Design 3.1—Exploring Resonance and Self-Oscillation Using Pink Noise • 64
High-Pass Filter • 64
Bandpass Filter • 64
Notch Filter • 65
Shelving Filters • 66
Allpass Filter • 66
Time-Based Effects • 68
Low-Frequency Oscillator • 68
Sample-and-Hold • 68
Sync/Free/Retrigger • 69
Delay • 70
Keyboard Tracking • 71
Multiple LFOs? • 71
Envelope Generators • 71
Other LFO and EG Applications • 73

Voltage-Controlled Amplifier • 74
Control Voltage • 75
Modifiers: Special Effects • 76
  Distortion • 76
  Delay-Based Effects • 77
    Echo • 78
    Reverberation • 78
      Acoustic Reverb: Halls, Studios, Cathedrals/Churches, Chambers • 78
      Mechanical: Plates and Springs • 79
      Digital Reverbs: Algorithmic and Convolution • 81
    Phaser • 83
    Chorus and Flanger • 83
  Ring Modulation • 84
Tuning—Fine vs. Coarse • 85
Unison/Voice Stacking • 86
Portamento/Glissando • 86
Modulation Matrix • 87
Arpeggiator • 88
Pattern Sequencer • 88
Oscillator Sync • 88
Pitch Bend/Modulation Wheel • 89

Instruments • 90
  The Digital Hardware-Based Synthesizer • 90
  The Keyboard Controller • 91
  The Sound Module • 91
  Software Synthesizers • 92
  Hybrid Synthesizers • 94

The MIDI Standard • 95
  Structure of MIDI Data • 96
  A Binary World • 97
  The MIDI Channels • 97
  Type of MIDI Messages • 97
    Data Bytes • 98
  Description of MIDI Messages • 98
  MIDI Connections • 103
  Beyond MIDI: Open Sound Control • 103

The Electronic Digital Studio for the Contemporary Composer and Producer • 104
  The Computer • 104
  The Audio Interface • 105
  The Controller • 105
  Monitoring System • 110
  Microphones and Preamplifiers • 110
    Microphone Types • 112
  Synthesizers • 112
  Portable vs. Home Studios • 114

4 Subtractive Synthesis • 115

Introduction • 115
Sculpting Sound With Subtractive Synthesis • 115
    Fundamentals of Working With Subtractive Synthesis • 117
Sound Design 4.1—Synth Clarinet • 119
Sound Design 4.2—1980s Classic Synth Strings • 120
Sound Design 4.3—Synth Snare • 121
The Producer Point of View on Subtractive Synthesis • 122
    Subtractive Synthesis Hardware and Software Sources and Options • 124
    Which Subtractive Device or Software Synthesizer is Right for My Productions? • 125
    Sonic Categories and Subtractive Patches • 127
  Bass • 128
    Sequencing and Production Techniques for Analog Bass • 128
    Mixing the Analog Bass • 130
  Pads • 131
    Sequencing and Production Techniques for Analog Pads • 134
    Mixing Subtractive Pads • 134
    Strings and Brass • 136
  Leads • 137
    Sequencing and Production Techniques for Subtractive Leads • 138
    Mixing the Subtractive Leads • 139
Exercises • 140

5 Frequency Modulation Synthesis • 141

Introduction • 141
How FM works • 141
Algorithms • 143
Operators • 144
Operators and the Harmonic Series • 145
Principles of FM Sound Design • 146
Sound Design 5.1—Constructing Sawtooth Wave in FM8 • 146
Sound Design 5.2—Square Wave • 147
Sound Design 5.3—Reed Pipe Organ/Oboe Timbre Using FM8 • 148
Modulation Index • 149
Amplitude of Sidebands • 150
Multiple Carriers • 150
Noise, Distortion, and Feedback! • 150
  Multiple Modulators • 151
  Key Sync/Key Tracking • 151
The Producer's Point of View on FM Synthesizers • 151
  Typical FM Synthesis Patches and Sonic Characteristics • 152
  FM Synthesis Hardware and Software Sources and Options • 153
  Sonic Categories and FM Patches • 154
  Bass • 155
    Mixing the FM Bass • 158
  Electric Pianos • 159

Mixing the FM Electric Piano • 160
Mallets • 163
Mixing the FM Mallets • 164
Pads • 166
Mixing the FM Pads • 168
Organs • 168
Exercises • 173

6 Additive Synthesis • 175

Introduction • 175
Instruments • 177
Resynthesis in Additive Synthesis • 178
Manipulating Resynthesized Sounds • 179
Spectral Synthesis • 180
Iris: Is it Sampling? Subtractive Synthesis? Spectral Synthesis? Yes • 181
The Producer Point of View on Additive Synthesis • 182
Sonic Categories and Additive Patches • 184
Pads • 185
Harmonics' Envelope • 189
Harmonics' Pitch • 190
Harmonics' Pan • 190
Adding Other Sources • 192
To the Next Level With the Use of Effects • 192
The Basic Patch • 194
High Impact and Direct • 194
Liquid • 197
Distant and Ethereal • 197
Disruptive and Aggressive • 201
Exercises • 202

7 Sample-Based Synthesis • 203

Introduction • 203
Principles of Sample-Based Synthesis • 205
Techniques for Custom Sample Sets • 205
Edit Mapping • 205
Pitch • 207
Groups • 207
Looping • 209
Release Samples and Control • 210
Being Creative • 210
Sound Design 7.1—Sampled Piano Manipulation • 211
Digital Audio: What the Heck Is Sample Rate and What Are Bits? • 211
Digital Audio in Practice • 213
File Sizes • 215
The Producer Point of View on Sampling • 215
The Sampler for Music Production and for Sound Creation • 216
Creating an Acoustic Drum Library • 217

Understanding, Recording, and Managing Multisamples • 219
    Preproduction • 221
    Programming a Drum Library • 225
    Mapping • 226
    Fine-Tuning the Samples • 233
    Working With Groups • 233
Programming Envelopes and Modulations • 235
Adding Effects • 238
Creating Variations Through Round-Robins and Random Sample Selection • 240
    Random Sample Variation in Kontakt • 241
    Round-Robin Variations in Kontakt • 243
Tips on How to Sequence for Acoustic Drums • 243
Creating a Beatbox Drum Kit • 246
    Editing the Samples • 249
    Sample Parameters and Effects • 249
    Creating Beatbox Patterns • 252
Working With Pitched and Sustained Samples • 253
    Recording the Samples • 253
    Programming • 256
    Looping • 256
    Programming • 259
Exercises • 261

8 Physical Modeling • 263

Introduction • 263
Analog Modeling • 263
Physical Modeling in Hardware • 264
Physical Modeling in Software • 264
Platypus of Sound? • 267
Logic's Sculpture as a Case Study • 267
The Producer Point of View • 268
    Some of Our Favorite Physical Modeling Synthesizers • 268
    Main Typologies of Physical Modeling Sounds • 269
    Sound Design and Production Techniques for PM Patches • 269
        Percussive Pluck Patches • 269
        Compositional Considerations • 274
        Pads • 275
        Ambient and Cinematic • 279
        Working on the Second Layer • 284
        Effects and Final Touches • 288
Exercises • 290

9 Wavetable Synthesis and Granular Synthesis • 291

Introduction • 291
Wavetable Synthesis • 291
    Wavetable Types • 292
        Lookup Table • 292

       Vector Synthesis • 293
       Linear Arithmetic Synthesis • 294
       "Advanced" Wavetable Synthesis • 297
Granular Synthesis • 298
The Producer Point of View on Wavetable Synthesis and Granular Synthesis • 300
    Wavetable Software Synthesizers for the Contemporary Producer • 301
   Creating an Aggressive Pad Patch With a Wavetable Synthesizer • 301
       Beginning and End • 305
       Balancing • 306
       A Bit of Filtering • 307
       Modulation Matrix • 308
       Effects • 309
       Rhythmic Pads With the Arpeggiator • 309
       Electronic Bass Using Wavetable Synthesis • 310
       Combining Arpeggiator and Modulation • 311
       Audio Effects • 312
   Granular Software Synthesizers for the Contemporary Producer • 314
       Real Time vs. Synthesizer • 314
       Granular Software Synthesizers • 316
       Building Granular Synthesizer Pads • 318
       First Generator—Layer A • 319
       Second Generator—Layer B • 320
       Envelopes • 321
       Modulation • 321
       Effects • 322
       Variations • 324
Exercises • 324

Index • 325

ACKNOWLEDGMENTS

This book has been a great collaborative effort with my dear friend and colleague Scott B. Metcalfe. His passion, knowledge, and enthusiasm have made writing this book an incredibly fun journey. This work would have not been accomplished without the fantastic support of my dear wife Irache and my wonderful daughter Alessandra. Their presence and love make my life an ever-creative sound design experience! I want to also thank my parents, my brothers Luca and Marco, and my dear friend Nella for their constant encouragement and support. Thank you!

Special thanks go to my colleagues and students at Berklee College of Music. I am very proud of being part of such a creative and talented teaching environment. In particular I would like to thank Provost Dr. Larry Simpson, Dean Dr. Kari Juusela, and Vice President Matthew Nicholl for supporting my sabbatical project, which has made writing this book a truly wonderful experience.

And finally, big thanks to Norm Hirschy and all the production and editing teams at Oxford University Press for having made such an enormous task fun and exciting.

—*Andrea Pejrolo*

Big thanks to my friend and coauthor Andrea Pejrolo from whom I have learned so much, and who has been a source of inspiration for many years. Such a pleasure working with you and exploring this topic that is near and dear to both of our hearts.

Thanks to my amazing and talented wife Sara for her support and patience through this sometimes arduous process and to my wonderful boys, Nathaniel and Andrew.

The research needed for this book could not have been possible without all the many helpful conversations over the years with colleagues and friends in and out of the music field: most notably David Budries, my longtime friend, teacher, and mentor; Dr. Robert Carl for his mentorship in composition; Dr. David Revill for his friendship and conversations on electronic music's evolution; Dr. Richard O'Brien for his help untangling some of the mathematics I encountered along the way; Ed Tetreault for bouncing around ideas with me and helping me gain a better understanding of the Hammond B3; Dr. Thomas Schuttenhelm for his encouragement and inspiring conversations about music; and my high school music teacher Scott Porter for showing me MIDI for the first time back in 1988 on an Apple IIc!

Big thanks to our editor at Oxford University Press, Norman Hirschy, for his vision and encouragement during this journey; and to Dave Smith and Joanne McGowan at Dave Smith Instruments; Thomas Dolby; Marc-Pierre Verge at Applied Acoustics Systems; Jimmy Bralower; Vincent Gutman; Larry Alexander; Morgan Walker at Korg; Arne Barlindhaug Ellingsen at Clavia DMI; Ian Vigstedt, Emmy Parker and Steve Maass at Moog Music; Tom Polk; Thomas Bloch; Dan Wilson; Dave Rossum, and Marco Alpert at Rossum-Electro; Peter Vogel; Ken Steen; Edwin Huet; and the many students and colleagues over the years at Peabody, Yale, and The Hartt School with whom I have had the privilege to explore the wonderful world of music technology.

—*Scott B. Metcalfe*

INTRODUCTION

Creating Sounds from Scratch was born out of the need for a resource that provides an understanding of how original sounds are created through the components that are common to synthesizers, *and* how they can be further manipulated for contemporary music production. Music artists use synthesizers all the time but stick to presets and *maybe* adjust effects here and there. It is our intention to break down the complexity of these powerful instruments to make them more accessible through easy-to- understand descriptions and practical examples.

An important component of this book is the "Sound Designs" through which readers are walked step-by-step through the process of building a common sound through the most common forms of synthesis used today. With each Sound Design, the barriers of your understanding will be eliminated so that you can unleash your creativity and start designing without getting lost in menus or aimless knob turning. You will learn how each type of synthesis can be used in specific styles and genres for contemporary music production.

Listeners are compelled by new and unique sounds; to use an example from popular music, it is as important as a "hook" or strong melody in catching a listener's attention. It is often easier for a contemporary writer/producer to use mainly presets, but the emotional impact of an original sonority in any type of music production (film, music, theatrical sound design, etc.) is far greater than using something that is familiar.

Any artists who use sound as their creative medium must be able to create something new and interesting to be successful. Electronic sound and music artists are no exception, and *Creating Sounds from Scratch* is intended to help you to produce the "right" sound and achieve your creative goals.

Making sounds from scratch is fun and (warning) more than a little addicting. Whether you are a professional or a hobbyist, we sincerely hope you find this collaborative effort to be a valuable resource in your ongoing development as a creative artist.

ABOUT THE COMPANION WEBSITE

http://www.oup.com/us/creatingsounds

Included on the companion website is video and audio content referenced in *Creating Sounds from Scratch*. We encourage readers to explore this material and revisit periodically as new content will be added indefinitely.

Creating Sounds from Scratch

1

How Did We Get Here?

Introduction

There is a pretty good chance you did not buy this book to learn about the history of synthesis and that you are more interested in jumping right in to make some sounds. We completely understand! In fact, there is no reason that you can't just skip ahead to Chapters 3–9 to get started learning how to make sounds with synthesizers. However, we strongly encourage you, now or after you have explored the rest of the book, to read through the first chapter to gain an understanding of the evolution of both hardware and software instruments in use today, and the second chapter, where we cover the physical and psychological aspects of sound. At minimum, the knowledge that follows in this chapter puts into perspective how your instrument evolved over a century of invention and refinement.

Figure 1.1
Composer Edwin Huet working the Moog modular synthesizer (photo credit: Metcalfe).

Creating Sounds from Scratch

The slippery slope of synthesizer obsession tends to follow this path: (1) You play a keyboard, run through its presets, and get really excited. (2) You get bored with those presets and start fiddling with knobs that say things like LFO and EG and Chorus and Reverb, perhaps having just a cursory understanding of its function—after all, what harm could be done? (3) And then you find yourself wanting to imitate a sound heard on a recording or develop something unique from scratch that you are hearing in your mind's ear and won't quit until you have it exactly. If this scenario sounds familiar, then welcome to the world of designing sounds from scratch!

The common controls on modern instruments may be familiar if not thoroughly understood because they are often "hardwired" under the hood (or bonnet, for our friends in the UK). "Synthesists" (if we may call ourselves by that name) who began their exploration of synthesis with modular devices were required to make patches with physical (hardware) or virtual (software) patch cords between components to manipulate sound. For musicians who benefit from a kinesthetic form of learning, this process is invaluable when later approaching software or hardware with perhaps graphically more elegant interfaces but obscured routing.

A good example of such a control is the the low-frequency oscillator (LFO), commonly used to create a vibrato effect: Level controls the intensity of the vibrato and Rate controls its speed. Someone working on a modular hardware or software system (see Figure 1.2) will accomplish the effect by patching the output of the LFO to the input of a voltage-controlled oscillator's (VCO) pitch modulation. The creative urge might lead to patching it elsewhere to see what might happen: How about the low-pass filter (LPF) that cuts high frequencies above a certain frequency? Ahh, regularly varying the brightness of the sound over time in a tremolo-like fashion. Lots of possibilities waiting to be explored.

Compare those very same capabilities seen clearly on a modular device with the Jupiter 8V (Figure 1.3). No question that the Jupiter is a powerful and wonderful

Figure 1.2
The clarity of signal flow when patches are routed physically or with virtual cables.

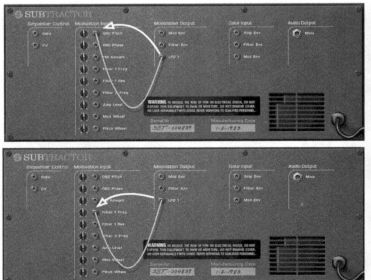

Reason Subtractor CV Patching   Moog Modular CV Patching

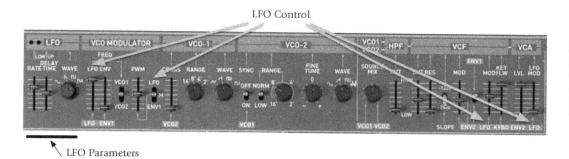

Figure 1.3
More common but less intuitive interface controls of the Arturia Jupiter 8V (emulation of the Roland Jupiter 8).

sounding synth in its own right, but the routing of the LFO to the VCOs or the VCF (or voltage-controlled filter; another name for a LPF) is less intuitive and may be intimidating to an uninformed user.

To summarize, many closed, nonmodular systems have these same capabilities but they may not be immediately apparent. Hardware digital synthesizers can be downright intimidating to the uninitiated because of the small liquid-crystal display (LCD) (relative to the comparatively large screen on a computer) and the often—shall we say—less-than-useful owner's manual. When confronted with this kind of interface, visualizing the routing can be very helpful.

Electronic Music in the Twentieth Century

Before we get too deep into the instruments themselves, let's first look at the artistic curiosities of composers in the early to mid twentieth century that led to the idea of creating new sounds for the first time with nonacoustic instruments.

As you may suspect, electronic music was born out of recording technology; more specifically, capturing and manipulating sound on audiotape. Edison cylinders and Emile Berliner's discs (a precursor to vinyl LPs) were the earliest formats for commercial music distribution, but did not lend themselves to creative uses beyond the capture of an acoustic performance. Things changed in a big way, however, when audiotape entered the picture, and it wasn't long before composers recognized its potential with new subgenres of tape-based composition such as Musique Concrète.

Musique Concrète

Although not specifically a process of music synthesis, Musique Concrète is a style of electronic music production that is worth exploring here as it was, arguably, a forerunner to sampling. Perhaps the easiest way to introduce you to the compositional/sound design style of Musique Concrète is through a familiar Beatles' song, "Being for the Benefit of Mr. Kite." In it there is a distinct "circus" section (starting at 1:52) assembled by George Martin and John Lennon, who used a technique borrowed directly from the early days of Musique Concrète only a decade or so prior. In his autobiography *All You Need is Ears*, George Martin paraphrases a discussion with the song's engineer, Geoff Emerick:

> ### Excerpt from *All You Need is Ears*
>
> *I got together a lot of recordings of old Victorian steam organs—the type you hear playing on carousels at county fairs—playing all the traditional tunes, Sousa marches and so on. But I clearly couldn't use even a snatch of any of them that would be identifiable; so I dubbed a few of the records on to tape, gave it to the engineer and told him, "I'll take half a minute of that one, a minute and a half of that one, a minute of that one," and so on.*
>
> *"Then what do I do with them?" he asked.*
>
> *"You cut that tape up into little parcels about a foot long, and don't be too careful about the cuts" . . . Fling them up in the air . . ."*
>
> *"Now," I said, "pick them up in whatever order they come and stick them all back together again."*
>
> —Martin and Hornsby 1994

Although creative, successful, and very effective—not to mention a great moment in the Beatles' groundbreaking "Sgt. Pepper's Lonely Hearts Club Band" album—this technique wasn't new to the music world, only to popular music. In the late 1940s and 1950s, French broadcaster and engineer Pierre Schaeffer established the "Musique Concrète Group" to explore new sonic possibilities by using a process not unlike what is described by George Martin. In the first of five études written by Schaeffer, in 1948, called "Étude aux chemins de fer" (in English, "Railroad Study"), he assembled various sounds of a locomotive (whistles, clatter of the wheels on the tracks, brakes, etc.) recorded onto tape and layered them into a surprisingly musical texture. This recording (which can easily be found online) is the first example of what Schaeffer labeled as Musique Concrète. It wasn't until his collaborations with composer Pierre Henry that the style gained more artistic merit, beginning with their first collaboration, "Symphonie pour un homme seul" ("Symphony for One Man"), inspiring other top composers of that era, such as Karlheinz Stockhausen and Pierre Boulez, to dabble with the technique (Simms 1986) (see Figure 1.4).

So what is Musique Concrète? The French translates to English as "concrete music," meaning music from real, found objects; music made from "found" sounds or recordings of sounds common to everyday life. According to the *Harvard Dictionary of Music*, " . . . Its basic idea is to replace the traditional material of music (instrumental or vocal

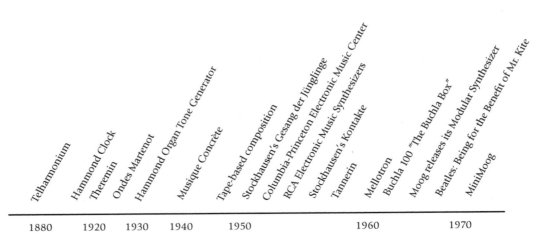

Figure 1.4
Timeline of electronic music developments prior to 1970.

sounds) with recorded sounds obtained from many different sources, such as noises, voice, percussion and others. As a rule this material is subjected to various modification: a recorded sound may be played backward, have its attack or resonance cut off, be reverberated in echo chambers, be varied in pitch." (Apel 1972).

A key element of the compositional style that would become standard after Schaeffer's first work was manipulating recorded sounds in such a way as to disguise their origin, thus altering the sound beyond recognition. A particularly effective technique, as suggested in the *Harvard Dictionary*, is clipping off the attack and/or resonance (analogous to "release" in the synthesizer world) portion of a sound's volume envelope. That may sound simplistic but it's quite an "ear"-opener, if you will, to take a very recognizable sound (piano, for example), cut off its percussive attack, and stop the sound before its natural decay is complete.

Audio Example 1.1[1]

Relating this to synthesis, listen to this example and hear how dramatically a piano sound sample is altered when performing the simple modification of cutting off the attack and decay.

It's interesting to note that these techniques are commonplace now with digital sampling, and George Martin's concern over any of the pieces of tape used in "Mr. Kite" being recognizable is shared by modern electronic musicians working in all styles of music—although staying within the parameters of "fair use" in today's copyright laws may be as much the objective as creativity.

Tape-Based Composition

During the same time period in New York, Vladimir Ussachevsky was at work exploring processes similar to those used by Schaeffer. "Tape music"—as it was commonly referred to—involved the use of analog recording tape to alter pitch by varying the speed on playback or feeding a piece of tape into the machine backwards so the sound would be reversed, and splicing for unrealistic but potentially interesting transitions. The source material was often recordings of real acoustic instruments that would be commonplace in a modern orchestra but represented here in a very different way—as with any good art, violating the listener's expectations.

In the early 1950s, Ussachevsky would begin collaborations with composer Otto Luening who took a similar approach with his work. The two presented a concert promoted by the American Composers' Alliance at the Museum of Modern Art in New York, a historic event followed by performances at a music festival in Paris and even a demonstration on NBC's *Today* show, bringing these new sounds out of the laboratory and into the ears of the general public.

---

1 Audio and video examples are available on the companion website.

Electronic Music Synthesis

The invention of the synthesizer cannot be attributed to any one person. However, there are important individuals who focused their understanding of electronics toward the generation of sound and artists with great foresight and creativity who imagined uses of the instrument beyond even the vision of its designer. Let's briefly explore some of this history.

It was largely the result of the collaboration between Ussachevsky and Luening that an electronic synthesizer would first come to the forefront of the electronic music world. In the mid-1950s, a confluence of events resulted in the founding of the Columbia-Princeton Electronic Music Center: (1) A need for more studio space led Ussachevsky and Luening to secure the use of facilities on the Columbia University campus; (2) RCA's demonstrating the "Olsen-Belar Sound Synthesizer" (later known as the RCA Mark II Electronic Music Synthesizer); (3) the pair securing a grant from the Rockefeller Foundation to purchase the RCA as a centerpiece for the studio; and (4) composer Milton Babbitt's interest in the instrument that led to his institution, Princeton University, joining in on the grant application. By 1959, the now-famous Columbia-Princeton Electronic Music Center was active and included an RCA Mark II and several tape studios. The 1960s would see a bifurcation of the electronic music tradition with one branch continuing on a path in the world of experimental/academic composition and the other toward popular music styles.

Karlheinz Stockhausen (whose early work was in the Musique Concrète style) was a pioneer of electronic music, writing what many consider to be masterpiece works of the genre, such as "Gesang der Jünglinge" and "Kontakte." The latter translates into English as "contact" and refers to the aural connection between the real instruments and electronic sounds in the piece. The first part of this two-part work ("Nr .12") is for electronics alone, and the second part ("Nr. 12 ½") adds piano and percussion. The former—which translates to "Song of the Youths," scored for electronics and recorded voices, merging tape-based composition with Musique Concrète—is considered by some musicologists to be, "perhaps the first masterpiece of electronic music" (Holmes, 2008). Writing for electronics allowed composers to expand beyond the limits of human performers and, as a result, benefited from the use of new kinds of musical notation.

In popular music, the synthesizer broadened the palette of songwriters and producers in similarly profound ways, providing colors that were new and intriguing to the ears of their audience. The experimentation of composers working on relatively primitive instruments led to the development of more advanced and compact designs and techniques that would become commonplace in the studio and on stage.

Instruments

Theremin

Where better to begin a discussion of early electronic instruments than with the Theremin (see Figure 1.5). Patented in 1928 by Russian Léon Theremin and originally known as an "etherphone" (because the performer plays the instrument without touching it), dynamics and pitch are controlled through the player's hands' proximity to a horizontally mounted loop on the left that controls amplitude (volume) and a vertical rod on the right that controls the frequency (pitch). As with many instruments, mastering the technique of playing a Theremin requires a great deal of practice to maintain proper pitch and control of the manual volume envelope. Making it even more of a challenge, having no physical contact with the instrument demands a discerning sense of pitch and precision to play in tune, which may be why it is better known for sci-fi sound effects than for musical performances.

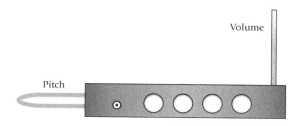

Figure 1.5
The Theremin.

Potentiometers (aka, "knobs") on the modern versions of the Theremin offer control over volume, frequency range, waveform, and brightness (cutoff frequency). The musical sound palette ranges from what might be described as violin-like to female-voice-like. Like that of nonfretted string instruments, vibrato is idiomatic to the Theremin and used to reduce the challenge of holding a steady pitch. It is created by the performer waving a hand or just a finger back and forth near the vertical rod.

The difficulty in playing a Theremin in tune led to adaptations of the original design that are heard on classic songs like "Good Vibrations" by the Beach Boys and "Echoes" by Pink Floyd; on modern songs from The White Stripes, David Gray and Guster; and eerie portamentos idiomatic of soundtracks of classic science fiction films. Among the most successful variations was trombonist Paul Tanner's "Electro-Theremin" (sometimes called a "Tannerin") and the Ondes Martenot that are each essentially a Theremin with a keyboard reference (see Figure 1.6).

Theremins continue to be manufactured by Moog Music, Inc. Until recently they remained largely unchanged. New models, however, incorporate what Moog refers to as a *heterodyning* oscillator that offers the performer the option of pitch correction ranging from strict semitones to full portamento (see Figure 1.7). And as you might expect, the Theremin, Tannerin, and Ondes Martenot all exist in software emulation, most notably on the iOS E-Theremin app.

Figure 1.6
Electro-Theremin or "Tannerin" uses a slide in front of a keyboard image to assist in locating pitches [left, photo courtesy of Tom Polk (http://www.tompolk.com)]. Ondes Martenot uses either a slide (in front of the keys) or the actual keys, depending on which mode is selected [right, photo courtesy of Thomas Bloch (http://www.thomasbloch.net)].

Figure 1.7
Moog "Theremini," which includes a heterodyning oscillator for optional intonation control (photo courtesy of Moog Music Inc.).

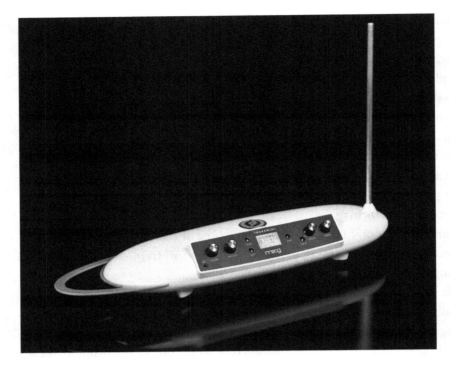

Hammond Organ

For the purposes of this book, we will be looking primarily at purely electronic instruments with two notable electromechanical exceptions: the Hammond organ and the Mellotron. We chose to include the Hammond because it can arguably be considered the first big step toward an instrument that was intended to emulate other sounds, and the Mellotron was significant in its introducing the idea of reproducing a sound from triggering prerecorded sources. There were many notable organs (like the Vox and Farfisa) that deserve mention as well, but it's hard to look past the success and longevity of the Hammond organ, particularly the B3 model.

The Hammond organ was the first mass-produced instrument that bridged the gap in a significant way between mechanical and electronic instruments. Although it had its predecessors—most notably the 1890s-era organ-like Telharmonium. which weighed in at a massive 200 tons, had 2000 electric switches(!), and transmitted performances over telegraph wires—the Hammond organ's unique design and rich sound quality have allowed it to find a home in many studios and on many stages alongside synthesizers. Its inclusion in the book points to its sound generation being directly comparable to additive synthesis, a process that we explore in detail in Chapter 6 (see Figure 1.8).

The genesis of Laurens Hammond's instrument began in the 1920s with a concept to build an electric motor. The first application of this motor was not for a musical instrument but rather for a highly accurate clock that would synchronize to its incoming mains a power frequency of 60 Hz (the standard in North America). The clock design was very effective at maintaining precise time. Hammond sought other uses for the same motor and saw its potential for providing an accurate pitch for an electric organ. It was in the 1930s when he patented a design that incorporated his synchronous motor into the organ's "Tone Generator," the electromechanical device from which the organ creates sounds.

Figure 1.8
Hammond B3 with Leslie speaker (courtesy of The Peabody Institute Recording Studio, Johns Hopkins University. Photo credit: Metcalfe).

The Hammond's Tone Generator uses ninety-one rotating tone wheels, each a small disc with protrusions (or bumps) around its circumference (see Figure 1.9). The bumps on the wheel cause the edge of the disc to continuously move closer to and farther from the magnet and coil as it spins. The result is a disruption of the magnetic field and induction of an alternating electrical current in the coil (not unlike the vibrating strings of an electric guitar running alongside pickups). The sound produced is something pretty close to a sine wave, which is what makes it a quasi-additive synthesis instrument. All ninety-one tone wheels move at a rate of 60 Hz (matched to the incoming power, of course). Tone wheels responsible for higher frequencies simply have a greater density of these bumps relative to those creating the lower frequencies and therefore create more waves per second.

Additive synthesis is all about building a timbre by adding the necessary harmonics. On the Hammond this is regulated with *drawbars*. Each drawbar controls the output level of a tone wheel generating as many as nine harmonics in much the same way you might assign multiple oscillators to produce sine waves in a pattern that mirrors the harmonic

Figure 1.9
Hammond tonewheel design.

Figure 1.10

Spectrum analyses of B3 organ drawbar settings.

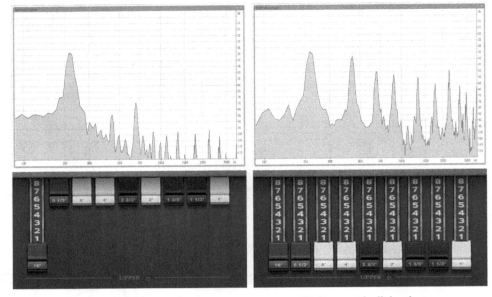

B3 Organ with fundamental only   B3 Organ with all drawbars open

series. In contrast with those of the additive synthesizer, however, the waveforms produced by the tone are not pure sine waves; they contain some subtle harmonic content that contributes to the rich quality associated with the instrument's character. Figure 1.10 shows the spectral contrast between only the fundamental drawbar at full volume versus all of the drawbars at full volume.

RCA Electronic Music Synthesizer

The first "programmable" (meaning it could not perform in real time) synthesizer was the RCA Electronic Music Synthesizer (see Figure 1.11), an enormous room-filling machine that looked more like a mainframe computer system of the time than a musical instrument. RCA engineers Herbert Belar and Harry Olson are credited with its design, with the first version of the instrument (the Mark I) installed at Princeton University in 1955 and, as already mentioned, the second (the Mark II) would become the centerpiece of the Columbia-Princeton Electronic Music Center in 1957.

Interestingly, the initial goal of Belar and Olson was not to create a composer's tool, but rather a device that would actually do the composing! In fact, they referred to it as an "electronic music composing machine" (Holmes 2008) with which they hoped to compose songs in the style of a given composer by generating notes somewhat at random within rules based on patterns found in a given composer's music. (An interesting idea but, to this day, attempts to imitate an individual composer's writing styles with technology have been less than convincing.)

The programmability of the RCA led the way to the sequencers we are familiar with today. Programming was accomplished with punch cards on which composers could specify tempo along with a pattern of notes to be played. Complex patterns and rhythms produced at fast speeds made the RCA desirable to composers intent on moving beyond the limitations of humans playing conventional instruments. Polyphony—the number of pitches a synthesizer can produce concurrently —was limited to four notes; a surprising

Figure 1.11
The RCA Mark II Synthesizer at the Columbia-Princeton Electronic Music Center at Columbia's Prentis Hall on West 125th Street, New York. Pictured: Milton Babbitt, Peter Mauzey, Vladimir Ussachevsky. "Ussachevsky Lab at 125th Street, 6/22/1959" (photo courtesy of Columbia University; photo credit: Manny Warman, photographer).

limitation by today's standards but a significant achievement for the time. Smaller instruments that would follow were monophonic, meaning they could play only one note at a time, well into the 1970s. The RCA altered the course of music composition, but its usefulness lasted only until the 1960s when the smaller, more user-friendly and less expensive designs became available.

The Rise of the Modern Synthesizer

In the twenty-first-century world of email and online user groups where developers and artists unrestricted by geographical distances collaborate on software and hardware design, it's difficult to imagine that development of the modern synthesizer is largely attributed to two individuals working separately in the same country, yet with no knowledge of the other's work. This was the scenario in the United States in the 1960s with Robert Moog working in Trumansburg, New York, and Don Buchla in Berkeley, California.

Moog

For musicians and lay people alike, one name comes to mind when discussing the synthesizer's foray into popular culture: Moog. On the East Coast, in upstate New York, an engineer and physicist by the name of Robert "Bob" Moog (see Figure 1.12) grew up in New York City and enjoyed building Theremins with his father at home. Armed

Creating Sounds from Scratch

Figure 1.12
Robert Moog with his instruments (photo courtesy of Roger Luther, Moog Archives. com).

with electrical engineering and engineering physics degrees from Columbia and Cornell, Moog focused his interests in music and electronics on building a synthesizer. His first commercial instrument was the "Modular" (see Figure 1.13). It was large by modern standards and somewhat complex to operate, but much smaller than the RCA, capable of real-time performance, and with a price tag that made purchasing one realistic for schools and well-off individuals. Its modular nature also meant that buyers could choose from a variety of à la carte components and build an instrument to suit their needs and budget. In 1969, prices were around $125 for an envelope generator, $395 per oscillator, and over $1000 for a sequencer.

Although brilliant when it came to engineering, Moog recognized his shortcomings as a businessman and musician and wisely sought the counsel of experts in those areas who would help refine his vision. Perhaps the most important advisor on the music side was composer Herbert Deutsch whose influence led to the use of a *keyboard* as the

Figure 1.13
Moog Modular synthesizer (courtesy of Computer Music department, Peabody Institute, Johns Hopkins University; photo credit: Metcalfe).

primary user interface. Arguably the biggest reason for Moog's success, the idea of including a keyboard was opposed by many composers of the era—Vladimir Ussachevsky, for one, felt strongly that the keyboard's semitone restriction would present an unnecessary limitation to modern composers who were beginning to break from the confines of equal tempered music (remember the interest in the RCA being capable of performing the humanly unplayable?). There are those who would go so far as to suggest that the synthesizer's becoming primarily a keyboard-based instrument was a setback in its development, although a strong case can also be made that Moog's success and the ubiquity of the keyboard as a synthesizer interface today is largely attributable to this one critical design decision. Even now, when synthesis controllers include guitars, woodwinds, brass, percussion, and other decidedly nonmusical instruments—like wireless batons and Nintendo Wii controllers—the keyboard still reins supreme as the de facto, tactile human–user interface.

The ubiquity of the name "Moog" in music lexicon and popular culture is due almost entirely to one recording: Wendy Carlos's monumental "Switched on Bach" album, on which Carlos successfully married old to new by performing Johann Sebastian Bach's music (written in the eighteenth century) on a Moog Modular synthesizer new to the twentieth century. Because the Modular is a monophonic instrument (capable of playing back only one note at a time), Carlos painstakingly recorded each voice part individually onto separate tracks of a multitrack tape for works ranging in complexity from the two-part Inventions to the *ten*-part Brandenburg Concerto No. 3. The album made it to the Top 10 on *Billboard*'s charts, became the first classical album to go platinum (selling over 500,000 copies), and took home three Grammy awards. A seminal work and a turning point in popular and serious concert music.

Synthesizers got smaller, becoming sufficiently portable for touring, and began to appear on concert stages with popular artists. Keyboard players had traditionally taken a secondary role among the rhythm section of rock bands playing piano or organ, but an amplified Moog could easily compete with drums and electric guitars, thus spotlighting for the first time rock keyboard players like Keith Emerson (of Emerson, Lake and Palmer), Rich Wakeman (of Yes), Richard Wright (of Pink Floyd) and Tony Banks (of Genesis), among many others.

Buchla

Meanwhile, on the West Coast of the United States, Don Buchla, completely unaware of Moog's pursuits until well into his own design process, was building an instrument of his own. Skilled with electronics and a strong interest in music (with experience as a composer and performer), Buchla's ambitions were to accommodate the needs of the composers with whom he was collaborating—most notably Pauline Oliveros, Morton Subotnick, and Ramón Sender (names among the most important American composers of the twentieth century). The Buchla Music Box Series 100 (see Figure 1.14) was designed to be a programmable sequencer-based instrument *without* a keyboard. Buchla's most significant contribution was his vision of a pattern sequencer, initially eight voltage steps (and later sixteen) that could be looped while other manipulations are performed live.

Creating Sounds from Scratch

Figure 1.14

Buchla 100 system (photo courtesy of Rick Smith, http://www.electricmusicbox.com).

It is important to note that the lack of a keyboard was not an oversight or miscalculation on Buchla's part but rather part of a design philosophy. Buchla was interested in human interaction with machines, and it was his opinion that the keyboard was a throwback to traditional design. He wanted his instrument to instead look forward with newly designed interfaces such as his "kinesthetic input ports," touch-sensitive metal pads housed in wooded boxes (Pinch and Trocco 2002). The user interface is where Buchla's and Moog's concepts obviously diverged most drastically, although other components, such as the modular construction and use of patch cords for customization, were notably similar solutions.

Commercial success was clearly on the mind of Bob Moog, and the keyboard made his instruments far more approachable to musicians who were less interested in programming. Even Vladimir Ussachevsky—who initially had approached Moog about buying some of his modules—instead chose to work with Buchla largely because of the keyboard issue. An interesting footnote, in 1973 Buchla released the Music Easel that included . . . wait for it . . . a keyboard!

Today, with vintage analog synthesizers very much back in vogue, Moog Music, Inc. continues to build Theremins and analog synthesizers, and Buchla Electronic Musical Instruments (BEMI) is back in business producing modular instruments very much in the spirit of their predecessors.

Mellotron

An essential tool for the electronic musician today is a sampler. If you assumed that sample playback was the sole domain of digital audio, you would be wrong! The 1960s saw the *tape*-based "sampler" known as the Mellotron—a bulky and, in some ways, impractical

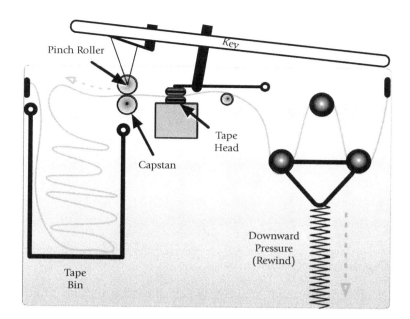

Figure 1.15
Diagram of the Mellotron's working mechanism.

design, yet one that made its way onto many majors recordings, particularly in the 1960s and 1970s. Perhaps its most famous use is the flute choir introduction to the Beatles' "Strawberry Fields Forever," but a quick search online will bring up lists of many hundreds of famous recordings that include Mellotron tracks.

So how did you play samples of recorded sounds before digital? By recording the sound to analog tape, and triggering playback by pressing a key that mechanically pulls through a fixed amount of tape by squeezing it between a pinch roller and capstan, and passing it over a tape head (see Figure 1.15). Although the description makes it sound relatively straightforward, building a successful device took a lot of mechanical engineering with many moving parts, thus making it very heavy (well over 300 pounds for early versions, and still over 100 pounds for later versions) and complex (translation: prone to mechanical breakdowns).

Unlike the modern sampler, the Mellotron had no practical way for users to record their own samples, but it offered composers and performers access to sounds that at the time were otherwise achievable only from the actual instruments being reproduced. The fixed tape length limited the duration of sustain, so performers would need to plan ahead when crafting a line. A convincing software emulation of the Mellotron will mimic this limitation.

The Mellotron was designed and initially sold by Harry Chamberlin in California. However, after a complicated and messy business deal (not uncommon in the music industry), it ended up being manufactured and distributed by Streetly Electronics in England. Chamberlin had begun his experiments on this technology way back in the 1940s and would continue to build instruments under his own name, although never achieving the level of success enjoyed by Streetly with the Mellotron.

The Road to Polyphony

To this point in our history timeline, with the notable exception of the RCA, synthesizers are still monophonic. The term *true polyphony* means an instrument with the means to produce multiple notes simultaneously and that each note has independent envelope control—a goal that was still out of reach to instrument designers of this era. Accomplishing true polyphony at the time required an organ-style design on which each semitone of the keyboard had its own oscillator, tone wheel, or pipe. The limitation in analog circuitry is the lack of a means by which note information from an individual key on the keyboard is sent to an available oscillator and envelope. On a multi-oscillator instrument, oscillators could be tuned to reproduce an interval or chord while one key was held down, but there was no facility for routing separate notes to available oscillators in real time. What follows here is an explanation of two clever analog solutions that were developed leading up to the advent of digital polyphony.

Duophonic/Duophony

The first step was "duophony," which allowed two pitches to be played simultaneously . . . sort of. Some implementations used a highest- or lowest-note priority, meaning the highest or lowest note would sound regardless of how many keys were pressed. The engineers at ARP (named for founder Alan R. Perlman) came up with something different for their Odyssey synth (see Figure 1.16), finding a way to send the pitch signal from the highest *and* lowest held keys and sending them each to a separate oscillator. So, if the player held down a C major chord (C–E–G) the C would go to Oscillator #1, the G would go to Oscillator #2, and the E would be ignored entirely. This design enabled a performance style with a sustained note in the left hand and a lead line in the right. However, because each note could use only one of the available oscillators, timbral options were limited, and if only one key was held, both oscillators would be layered on the same pitch, thus changing the resulting timbre.

Audio Example 1.2

Demonstration of Duophonic Playing Techniques

Figure 1.16
Modern Duophonic Instruments: ARP Odyssey (left, courtesy of Korg USA Inc.) and Moog Sub 37 (right, photo courtesy of Moog Music Inc.).

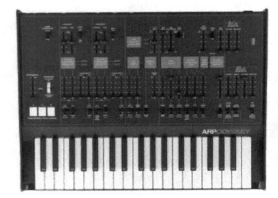

Paraphony/Paraphonic

The mid-1970s saw another solution called "paraphonic." A paraphonic synthesizer allowed several notes (four or eight was common) to sound simultaneously, but all would share the same filter and/or envelope generator—meaning that as soon as the first note began, the envelope shape would begin as well; any subsequent notes would join the envelope shape in progress rather than starting from the beginning with its own envelope shape (see Figure 1.17). This is a little confusing, but consider the example of a real piano: As each note is struck, it follows an envelope of fast attack and long decay, regardless of whether the notes are hit concurrently or in sequence. With a paraphonic synth, several notes can sound concurrently but the envelope and filter shape are all tied to the first and last notes that were played.

The paraphonic design was useful for pad sounds, but its limitations could be frustrating. It was becoming clear that the only means for achieving true polyphony with an analog synth was by dedicating an oscillator (or oscillators), a filter, and an envelope generator to *each* key—big, expensive, and impractical. Moog's attempt at this design was the Polymoog which was big (over 80 pounds) and expensive (over $5,000US in the 1970s), and its complexity made it unreliable and, thus, impractical (see Figure 1.18). However, it did offer true polyphony: capable of sounding all of its seventy-one keys simultaneously, each with a dedicated filter and envelope.

Figure 1.17
Graphical illustration of monophonic, paraphonic, and polyphonic envelope shapes.

Figure 1.18
Polymoog (photo courtesy of Moog Music Inc.).

> It is worth noting another early innovation from Laurens Hammond that falls into the polyphonic category as well: the Novachord, from *1939!* Well ahead of its time in many ways, the Novachord was a beast weighing in at 500 pounds made with beautiful wooden cabinetry, 163 vacuum tubes, more than 1,000 capacitors, and lots or wires. It was truly polyphonic: Each of the 72 lets had its own dedicated amplitude envelope. There was even a filter stage and an *electromechanical* LFO (see Figure 1.19).

Figure 1.19

Hammond Novachord (Photo courtesy of Dan Wilson, Hideaway Studio, UK).

Arrival of True Polyphony

It became clear that analog technology alone was unsuitable for accomplishing true polyphony. By some means, the synthesizer would need to gather information about which keys were being pressed when and assign them to an available oscillator, each with a dedicated filter and envelope. The number of simultaneous notes in such an instrument would be dictated by the number of oscillators installed. The only practical solution for the problem was to use a digital microprocessor.

With digital control of analog oscillators, the processor assigns a *voice* to a binary address (e.g., somewhere between 000000 and 111111). Each voice is a stand-alone monophonic synthesizer with its own dedicated oscillator, filter, and envelope. The keyboard is constantly scanned by the processor for "on" and "off" signals as keys are pressed and released). When a key is detected, the processor sends the control voltage for that note to an available voice and plays the sound at that pitch (private communication, Smith 2015). Sequential Circuits, led by Dave Smith, was the first to implement a microprocessor into an analog synthesizer in 1978 with their enormously successful Prophet 5 keyboard (see Figure 1.20).

Figure 1.20
Sequential Circuits Prophet 5.

Thomas Dolby on His Synthesizer Work on Foreigner's "Waiting for a Girl Like You"

"For the intro of [Foreigner's] 'Waiting For A Girl Like You' I used a Minimoog to record several sustained single notes in a minor scale, each on its own track of a 2" multitrack tape. I then "played" the faders on the mixing board in the back room at Electric Ladyland (a 10 channel Neve 8068-era sidecar) as if I was playing a Mellotron, and bounced the result down to two of the tracks. When I first played it to the band, the drummer described it as "massage music." But Mick Jones liked it and it stuck!

"For the melody (once the drums come in) I doubled a Prophet 5 with a Roland JP4. I made the sounds as similar as I could and played one with each hand. This was before MIDI [musical instrument digital interface]. Mutt Lange made me do about a hundred takes till I was ready to drop. He has the most microscopic hearing. 'That was great!', he would say. 'Although the fifth note of the third bar was a smidge rushed.' 'So shall we punch in for it?' I asked hopefully. 'Nah, let's just to the whole thing over.' This went on for hours."—Thomas Dolby (January 2015)

Musical Instrument Digital Interface

It was Dave Smith, founder of Sequential Circuits and designer of the Prophet 5, who in 1981 first proposed the idea of a "Universal Synthesizer Interface" in a paper presented at the Audio Engineering Society Convention of 1981 (private communication, Smith 1981). The dream was for an interface to be adopted by all manufacturers that would eliminate the proprietary controls for real-time performance information that companies were introducing independently.

Already in existence but limited in functionality was the somewhat universal "control voltage" through which envelope shape or LFO, for example, from one instrument, could be sent out and applied to a filter or voltage-controlled amplifier (VCA) on another. Manufacturers like Oberheim and Roland were independently working on systems that could transmit and receive note-on and note-off, pitch, velocity (how hard a key is struck), program change, and other information from a "controller" keyboard and send it to another "slave" keyboard. An obvious application would be recording a performance in a digital player-piano-like fashion, but unlike that with analog or digital audio recording, the tempo could easily be altered without affecting pitch, and wrong notes could be corrected with ease. New timbres could be designed by the layering of a patch from a "master" keyboard while the "slave" keyboard followed right along with its own patch. Such designs were in use on instruments by these companies but they were only compatible within their own product lines: Roland had a system called "DCB" (digital control bus), and Oberheim had its own called the "Oberheim Parallel Bus."

> According to Dave Smith, "... Sequential had a high-speed serial interface prior to MIDI that was actually 10 times faster than MIDI. I believe Yamaha may have also had an in-house system. That was why MIDI was instigated; there were too many non-compatible interfaces. MIDI became ubiquitous since it was royalty-free, easy and inexpensive to implement for manufacturers, and provided functionality for 95% of musician's requirements, while being simple enough to understand" (private communication, Smith 2015).

Smith proposed that manufacturers bring ideas for a standard protocol to the 1982 National Association of Music Merchants (NAMM) conference. Not many showed up, but Ikutaro Kakehashi of Roland did, and he agreed to work with Smith on a solution. Subsequently, the 1983 NAMM show became one for the history books (see Figure 1.21). For the first time ever, two instruments from otherwise competing manufacturers: Roland with their Jupiter 6 and Sequential Circuits with their Prophet 600—each instrument included three five-pin "DIN" ports labeled *MIDI* (Musical Instrument Digital Interface) IN, OUT, and THRU. Connect the OUT of one instrument to the IN of the other, and voila! ... two synthesizers playing back simultaneously from one keyboard.

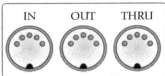

Figure 1.21
A photo from Dave Smith's personal collection from NAMM 1983 when a Sequential Circuits Prophet 600 and a Roland Jupiter 6 were connected via MIDI, the first time a performance communication was transmitted between two devices made by different manufacturers (photo used by permission of Dave Smith).

Smith decided to make the MIDI protocol available for free to any company that included it on their instruments. That bit of rare generosity caused MIDI to spread to the point of ubiquity, so much so that a keyboard manufacturer now would not even think of releasing a new product without MIDI being fully implemented. New instruments still have the exact same DIN-5 connectors as they did in 1983, but MIDI can also be transmitted and received through universal serial bus (USB) and FireWire (IEEE 1394) connections, making multichannel two-way communications with a controller and computer-based software as simple as a single-cable connection.

> MIDI communication is not limited to instruments and computers; it is also used for automation of audio mixers and stage-lighting consoles. Some cellphones even use MIDI for ringtones, a technology originally developed by Thomas Dolby's company Headspace (later named Beatnik) and licensed to Nokia.

Digitally Controlled Oscillator

Synthesizers were evolving quickly as the 1970s turned into the 1980s. Digitally controlled analog instruments offered storage of recallable presets and polyphony, but the analog oscillators were unstable and intonation consistency was a problem. Digitally controlled oscillators (or DCOs), developed by Philips Corp. in 1969 and implemented in electronic organs in the 1970s, began to appear on large and expensive synthesizers like the Synclavier in 1975 and the Fairlight CMI (Computer Music Instrument) that came out in 1979.

Synclavier

Heading back to the 1970s for just a moment to look at the beginnings of digital synthesis, Jon Appleton, a music professor and experimental composer at Dartmouth College, was looking for a way to eliminate the inconvenience of all the patching that was necessary on the modular synthesizers by utilizing a newly installed time-sharing computer that could be accessed from terminals around the campus (Dartmouth was the first school to incorporate such a system). Along with Sydney Alonso, an engineer and digital electronics specialist at Dartmouth, he researched the possibility but determined that the power needed to address their needs would overburden the system. However, they thought that it might be possible to build a fully digital synthesizer. Along with Cameron Jones, an engineering and music student who studied with them both, they put together a simplistic system for the time—at least in terms of its capabilities—but one that would foreshadow the future of *digital* music making.

These experiments would eventually result in a commercial product in 1978, the Synclavier, produced by New England Digital (a company founded by Jones and Alonso), and sold primarily to universities. It was in 1980, however, with the release of the Synclavier II, that its popularity started to really take off, despite its price tag of between $31,000 and $71,000, depending on the configuration. The system had evolved from a simple additive and frequency-modulation (FM) synthesizer to include a sampling option (for a mere $30,000 when trading in your sound modules) and the beginnings of what would become wavetable synthesis. Its capabilities as a 16-bit, 100-kHz sampler with thirty-two voices and stereo panning put it in a league of its own, attracting top-selling artists such as Michael Jackson, who used it on the "Thriller" album (most notably for the electronic gong sound at the beginning of "Beat It"). It also became a favorite of TV and film composers such as Patrick Gleeson, who used it to score *Apocalypse Now*, Mark Knopfler, for *The Princess Bride*, and Mark Snow, for *The X-Files*.

Chowning and the FM Synthesizer

Integral to any creative process is the question, "What would happen if I . . . ?" The results of such an inquiry fall somewhere on a scale of groundbreaking to "Oops, I won't do that again." Fortunately for those of us who work with sound, an "Oops" result rarely yields consequences beyond time spent learning what *not* to do next time. An unexpected result or one encountered by accident are gifts of creative exploration sometimes discarded as mistakes or something that is not terribly useful . . . or, on the other hand, something entirely new with enormous potential. It is all but certain that sonic adventurers stumbled on FM either intentionally or unintentionally, but moved on because of a lack of understanding of what it was they were hearing and how it could be useful as a means of sound design.

Fortunately, in 1971, composer and Stanford University Music Lecturer John Chowning came along and was determined to make sense of the "extreme vibrato" he heard when modulating one VCO with *another* VCO. It was standard practice, of course, to modulate the frequency of a VCO with a LFO, but modulating the pitch with *another* VCO was not recognized at the time as fruitful. Arranging VCOs (known in FM synthesis as *operators*) in different modulation configurations called *algorithms* form the raw material from which sounds are created in FM. Although modern FM instruments may include filters, they might be considered redundant because modulation control itself results in the presence and arrangement of harmonics in the sound produced. Envelope generators were used for overall amplitude shaping and could be applied to the modulation signals themselves as a way of shaping harmonic content over time.

Through collaboration with colleagues at Stanford, Chowning published what would become a seminal paper in music technology, "The Synthesis of Complex Audio Spectra by Means of Frequency Modulation" in the *Journal of the Audio Engineering Society*, September 1973. American companies—reportedly including Baldwin, Allen, and Hammond—had no interest in licensing the technology, but a chance meeting with engineer Kazukiyo Ishimura of the Yamaha Corporation led to a collaboration that would have Yamaha implementing the technology in the next decade into their DX-series keyboards, the first and most famous of which was the DX7 (see Figure 1.22) (Grunwald 1994).

Figure 1.22
The Yamaha DX7 II-FD (first stereo version in the DX series).

An important development to the history of FM synthesis was the aforementioned DCO. Although it is possible to do FM synthesis with analog oscillators, the stability and accuracy of the DCO allowed for much greater control and predictability. As you will discover when exploring Chapter 5, FM synthesis is totally different and a bit less intuitive than subtractive methods used in previous synthesizers. The precise, stable frequency offered by a DCO for both the *carrier* and *modulator* signals is necessary for predicable designing of sounds. This is not to say, however, that FM should be avoided with analog oscillators but the results vary and can be hard to reproduce later (something as simple as heat affects the frequency of analog oscillators, and the slightest variations pitch can have a dramatic effect on FM).

Digital Sampling

Synthesizer players loved their instruments because the sounds they could produce were fresh. Research has shown that, to a listener, compelling music is not just about a catchy melody or interesting rhythm, it's also important that its timbre be novel. If the goal, however, is to emulate a real acoustic instrument, the Mellotron had been the best available option. Digital technology had revolutionized synthesis by adding polyphony and pitch stabilization, and now the time was right to apply digital to playing back recorded sounds.

The first instruments to incorporate digital sampling were the Synclavier II (discussed earlier) from New England Digital and the Fairlight CMI Series I that came from Australia in 1979 (see Figure 1.23). In what began as an attempt at physical modeling, the CMI designers, Peter Vogel and Kim Ryrie, instead shifted their design to playing back recorded samples and immediately saw the potential.

Figure 1.23
Thomas Dolby with a Fairlight CMI (photo courtesy of Thomas Dolby).

Creating Sounds from Scratch

Figure 1.24
E-MU Emulator (photo courtesy of Dave Rossum and Marco Alpert).

Although early versions suffered in sound quality, the CMI Series III (released in the mid-1980s) provided 16 bits and a 44.1-kHz sampling frequency (the same as those of the compact disc). Somewhat more practical than the Synclavier in terms of size and cost (£20,000/$65,000US), the Fairlight found favor with top artists of the time, including Peter Gabriel, Thomas Dolby, Herbie Hancock, Stevie Wonder, and many others.

There is an important distinction between *samplers* and instruments that are simply *sample players*. Sample players, as the name implies, are preloaded by the manufacturer with a set of samples from which to choose for playback. Samplers can do the same but additionally offer the user the ability to record and edit their own samples. One of the first companies to offer full sampling capabilities in a compact and relatively affordable package was a company called E-MU Systems. Their Emulator series was enormously successful at bringing sampling to a much wider user base (see Figure 1.24).

We would also be remiss in our duties were we not to include a brief mention of the Linn LM-1 Drum Computer (see Figure 1.25) in this category. Its successor, the LinnDrum, is an example of an instrument that was a sample player with a very specific objective: to play samples of drums and allow programming of rhythm

Figure 1.25
Jimmy Bralower pictured with the Linn LM-1 Drum computer (photo courtesy of Vincent Gutman, Marc Inc of New York).

patterns. There were "drum machines" that predate its release in 1980, but the Linn was considered to be the first serious professional device in the category and one that would find its way onto the rhythm tracks of countless recordings. There were even drummer–producers who specialized in Linn programming, among the most notable being Jimmy Bralower.

> **Conversation With Producer–Drum Programmer Jimmy Bralower**
>
> The trend for drum sounds in the 1970s was moving toward greater isolation of the individual elements making up the kit. Microphones on each drum were fed to their own track on the then-new twenty-four-track tape machines. Studios were treated with lots of absorptive materials to minimize leakage between mics, and engineers used triggers on drums to open gates, effectively unmuting a mic when the closest drum or cymbal was hit. The disco style called for a four-on-the-floor kick drum that was in lock step with a click track. Opening a gated kick with a click track meant the beat would be "perfectly" in time. In short, session drummers were being asked to play more like machines and had to relearn their technique, which now put more emphasis on rigid timing, less so on subtle shifts that were part of a player's feel or the rhythm "pocket" they were forming.
>
> New York's Jimmy Bralower was immediately drawn to the possibilities offered by the LinnDrum. He saw how he could create and layer unique drum and percussion patterns, delivering the tight feel producers were now expecting, and, as an experienced studio drummer, making the results more musical by applying subtle techniques developed from years of playing real drums. One example could be detuning a snare lower; it might still fall exactly on beats two and four in a 4/4 bar, but the lower lower pitch makes it sound (appropriate for the feel) a hair late. Jimmy led the way with an approach and technique that achieved a level of artistry any great musician does with an instrument.
>
> In addition to the Linn, he also made great use of the Simmons, Roland's 808, the Linn 9000 (first with sampling and sequencing), the Akai MPC, and samples triggered on many instruments for recordings with Hall & Oates ("Big Bam Boom"), Steve Winwood ("Higher Love"), Laurie Anderson ("Strange Angels"), Madonna ("Like a Virgin"), Peter Gabriel ("So"), Cyndi Lauper ("True Colors"), Eric Clapton ("Journeyman"), and countless others starting in the 1980s.

Wavetable Synthesis

Between digital samplers and subtractive synthesizers lie wavetable synthesizers. Digital samplers are intended to reproduce a sound that ranges from a few to many seconds (or more) in length and provide some facility for manipulating that sound. Digital sampling requires fast access to the data required for playing back a recorded sample immediately on request (i.e., pressing a key on a keyboard), and is therefore usually stored entirely in random-access memory (RAM). When RAM was expensive and available only in much lower capacities than what we are accustomed to today, samples were shorter with low bit rates (such as 8 bit, compared with today's 16 or 24), low sample rates, and mono. An attractive alternative was storing and reproducing a single wave cycle that was very short

Figure 1.26

Wavetable ES2 and Reason's Malström wavetable instruments.

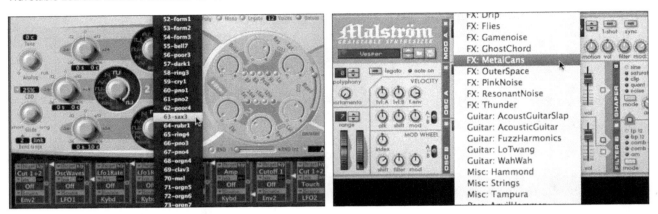

(in the range of milliseconds), therefore requiring minimal storage space and RAM, which is where wavetable comes in.

It was Appleton and Alonso at New England Digital who were looking to achieve more realistic and musical sounds within the limited capabilities of their computer who first thought to divide a sound event up into three small chunks—beginning, middle, and end—and store them in a "wavetable."

Working with single-cycle waveforms harkens back to the options of the early days of synthesis, but with a wavetable, the waves are far more complex. Rather than choosing from sine, triangle, square, and saw in a traditional synthesizer sound palette, the options now are expanded to a variety of raw sounds.

Wavetable synthesis is still found on synthesizers sold today, like those in Figure 1.26 and others that take the concept even further. For Logic users who have not carefully read their owner's manual this may come as a surprise, but Apple has hidden a list of 100 waves that can be selected from a table that pops up in the first oscillator. Reason's Malström is a hybrid instrument designed to incorporate aspects of wavetable synthesis with a variation known as granular synthesis (both explored in greater detail in Chapter 9).

Not unlike how Robert Moog adapted existing technology into a commercially viable product, digital synthesis would see its first step into the mainstream with Palm Products GmbH (PPG) founder Wolfgang Palm and the release of the PPG Wave series in the early 1980s (see Figure 1.27). It was his frustration with the limitation of analog

Figure 1.27

PPG Wave 3.V (Waldorf's software emulation of the PPG Wave synthesizer).

Figure 1.28
The Korg Wavestation (photo courtesy of Korg USA Inc.).

synthesizers (such as oscillators that would not stay in tune) that led Palm to explore the possibility of sound sources being stored as digital snippets offering pitch stability and new timbres.

Despite the expanded palette of sounds, early versions of The Wave also suffered from a low 8-bit quality and lacked common modifier features of its analog cousins. The second version, Wave 2, combined the digital sources with a resonant *analog* LPF, an LFO, and a variety of envelope shapes. This successful hybrid design was expensive enough (just under $10,000) to limit its users to those in the upper tier of the popular music world but served as inspiration for more affordable designs that came later from big companies like Roland, Korg, and Yamaha.

Another important hardware wavetable instrument in the 1980s was the Sequential Circuits Prophet VS, which provided the user a joystick for blending from four selectable wavetables. Although it was a great instrument (discussed in more detail in Chapter 9), sales were disappointing and the company was sold to Korg. Korg took the concept and ran with it in their Wavestation that was released in the early 1990s (see Figure 1.28). On it, wavetables could be reproduced in sequence, offering dynamic timbral shifts over time—a great tool for producing rich pads, for example.

Dynamically shifting through different wave shapes is standard for wavetable instruments. The ES2 and Malström both allow using a LFO and/or envelope generator to scroll through the list during a performance. Generally, a wavetable list is organized so that each step is a slight variation on what precedes and follows, thus facilitating a smooth transition from one to the next.

Korg M1 Workstation

By 1988 the the time had come to combine some of these emerging technologies into a single *workstation*: a keyboard synthesizer with built-in sound modules and an onboard MIDI sequencer for multitrack recording (see Figure 1.29). The M1 predates the Wavestation by a couple years but is worth mentioning here as it was hugely successful in this regard. Its sound palette was based on samples that could be manipulated with all the functionality of subtractive synthesis. Starting with samples gave the patches an organic quality that remained evident even when the sounds were vastly manipulated beyond the point where the origin was recognizable.

Creating Sounds from Scratch

Figure 1.29
The Korg M1 workstation (photo courtesy of Korg USA Inc.).

Conclusion

It is impossible to include all of the technologies and people and instruments that have made an impact in the evolution of synthesis, but those represented here are among the most important (see Figure 1.30). In Chapter 2, we will look at important concepts of sound as a physical entity and the sometimes curious human perception of sound to prepare you for exploiting both. Chapter 3 covers the hardware and software interfaces and parameters common to all forms of synthesis, and then the remainder of the book will focus on a different form of synthesis for each chapter. Buckle up, let's go for a ride.

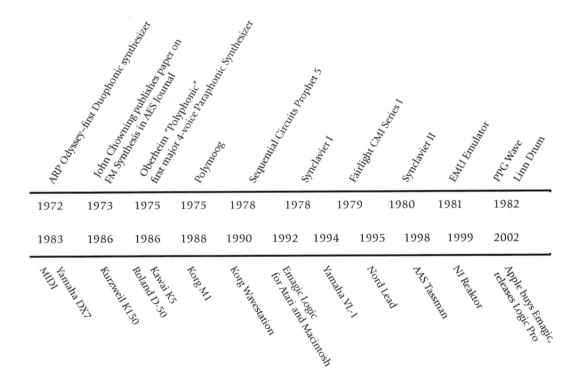

Figure 1.30
Timeline 1972–present.

Acknowledgments

Marco Alpert, Rossum Electro-Music LLC
Thomas Bloch [http://www.thomasbloch.net]
Jimmy Bralower [http://jimmybralower.com]
Thomas Dolby
Roger Luther [http://www.moogarchives.com]
Steve Maass, Emmy Parker, and Ian Vigstedt at Moog Music Inc.
Joanne McGowan, Dave Smith Instruments
Tom Polk [http://www.tompolk.com]
David Revill
Dave Rossum, Rossum Electro-Music LLC (cofounder of E-MU Systems)
Dave Smith, Dave Smith Instruments (founder of Sequential Circuits)
Ed Tetreault
Peter Vogel, Peter Vogel Instruments (cofounder of Fairlight Inc)
Morgan Walker, Korg USA Inc
Jocelyn K. Wilk, Columbia University Rare Book & Manuscript Library

Bibliography

1. Apel, Willi. *The Harvard Dictionary of Music*, 2nd ed. Cambridge, MA: Belknap Press, 1972.
2. Grunwald, Eric. *Bell Tolls for FM Patent, but Yamaha Sees "New Beginning."* Stanford Technology Brainstore, Vol. 3, No. 2, Summer 1994, Palo Alto, CA.
3. Holmes, Thom. *Electronic and Experimental Music*, 3rd ed. New York: Scribner, 2008.
4. Lehrman, Paul. Interview with John Chowning. *Mix Magazine*, March 2005.
5. Levitin, Daniel. *This Is Your Brain on Music*. New York: Dutton-Penguin, 2006.
6. Martin, George and Hornsby, Jeremy. *All You Need is Ears*. New York: St. Martin's Press, 1994 (reprint edition).
7. Metcalfe, Scott. Phone interview with Jimmy Bralower. October 9, 2015.
8. Metcalfe, Scott. Email interview with Thomas Dolby. January 22, 2015.
9. Metcalfe, Scott. Email interview with Dave Smith. February 2015.
10. Pinch, Trevor and Trocco, Frank. *Analog Days: The Invention and Impact of The Moog Synthesizer*. Cambridge, MA: Harvard University Press, 2002.
11. Russ, Martin. *Sound Synthesis and Sampling*. Oxford, UK: Focal Press, 2004.
12. Simms, Bryan R. *Music of the Twentieth Century, Style and Structure*. New York: Schirmer Books, 1986.
13. Smith, Dave and Wood, Chet. "The 'USI', or Universal Synthesizer Interface." Paper 1845, presented at the 70th Audio Engineering Society Convention in New York, October 1981.
14. Vail, Mark. *The Synthesizer*. New York: Oxford University Press, 2014.
15. Vail, Mark. *Vintage Synthesizers*. San Francisco: Miller Freeman Books, 2000.
16. Wagner, Kurt. "One of tech's most successful inventors never made a cent." *Fortune Magazine*, April 11, 2013. http://fortune.com/2013/04/11/one-of-techs-most-successful-inventors-never-made-a-cent/

2

Understanding Sound and Hearing

Introduction

Why devote an entire chapter of a book on synthesis to understanding sound and hearing? Unlike sculptors or painters who have the distinct advantage of being able to see and touch their work, the sound artist's medium is invisible. By understanding the physical aspects of sound and the psychoacoustics of how we interpret sound, we can begin to visualize what it is doing and have more precise control over its manipulation.

What Is Sound?

Vibration of air particles that occurs within an intensity and frequency range in which human ears are capable of sensing it results in the perception of sound. Too low intensity does not mean the absence of vibration, it is simply insufficient for us to hear; an intensity too great inflicts pain and/or permanent damage to our ears. That range is measured as 0.00002 to 2 Pa, or, more conveniently, as 0 to 120 dB-spl (decibels of sound pressure level) The audible frequency range is measured in hertz (Hz), the number of complete wave cycles that occur within a second of time, and is roughly 20 to 20,000 Hz (or 20 kHz). Factors lowering high-frequency sensitivity include damage from overexposure to normal age-related hearing loss.

A complete wave cycle is defined as a region of high pressure called compression immediately followed by a region of low pressure called rarefaction (or vice versa) (see Figure 2.1). Each complete cycle that occurs within 1 s of time is its frequency. The physical size of the wave is frequency dependent. The velocity of sound is the same regardless of frequency, so for 1000 Hz to travel the same distance as 100 Hz, for example, the wavelengths must be physically smaller.

The origin of a sound might be a musical vibration, like the reed of a clarinet or a player's lips on a trumpet mouthpiece, or nonmusical, like shoes

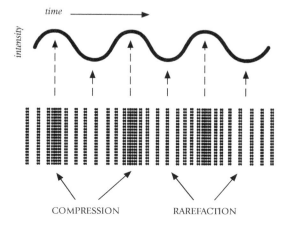

Figure 2.1
Sine wave shown (top) as a graph of its positive and negative intensity fluctuations over time, and (bottom) corresponding areas of compression and rarefaction where air particles are close together and farther apart, respectively.

Creating Sounds from Scratch

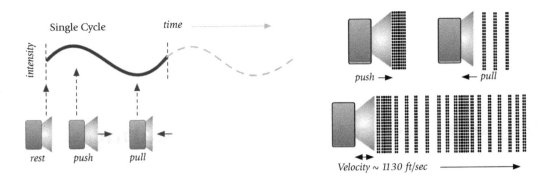

Figure 2.2
Areas of compression and rarefaction caused by the vibration of a loudspeaker driver.

striking a solid floor. In each of these scenarios, some of the energy in the physical motion is transferred as a disturbance in the surrounding air particles. Basic physics tells us energy cannot be created or destroyed, only its form changes: Physical energy is becoming acoustical energy in this scenario. Figure 2.2 shows this process with the example of a loudspeaker driver.

The smooth back-and-forth motion of the driver in this example produces a sine wave. Imagine attaching a pencil to the driver and, during playback, pulling a long piece of paper at a right angle to the motion of the driver. Assuming the intensity is great enough, you will see a sine wave taking shape on the paper. The intensity or amplitude is displayed by the height of the wave, equivalent to the distance of the driver's travel in and out.

The disturbance created in the air moves away from the source at a rate of approximately 1130 ft/s (or 344 m/s), depending on air temperature and other factors. That seems really fast until you compare it to the velocity of light which clocks in at 186,000 miles/s! That disparity is most apparent when an object generating a loud sound can be seen from a distance. In the case of a baseball game, for example, if you are seated far away from home plate, you will see the bat strike the ball before you hear the "crack." Even on a small scale the relative slow pace of sound demands the attention of a recording engineer with microphones close to, say, a snare drum, and a stereo pair of microphones over the entire kit. For easy numbers, if we round its velocity to 1000 ft/s we can say that it takes about a millisecond (ms) for sound to travel the distance of 1 ft. If the drum overheads are 3 ft from the close mics there will be a 3-ms delay. Doesn't sound like much, perhaps, but it's enough to smear the sound when the two are blended and is usually something the engineer will choose to address.

Timbre—Harmonic Structure

Timbre (pronounced: tám-bər) is defined as the distinct character or quality of a sound, independent of its pitch and intensity. An example would be the difference between a clarinet and trumpet each playing an A-440 pitch. The pitch is the same but the sounds are distinct. That distinction is caused by the presence of waves at frequencies other than the 440 Hz fundamental pitch.

Both instruments in Figure 2.3 are playing the same pitch, as is evident by the loudest harmonic at 440 Hz. Above the fundamental are harmonics, in this case positioned at integer multiples (2x, 3x, 4x, ...) of the fundamental frequency: 880 Hz, 1320 Hz,

Figure 2.3

Amplitude over frequency-spectrum analyses of a clarinet (left) and a trumpet (right) playing an A-440.

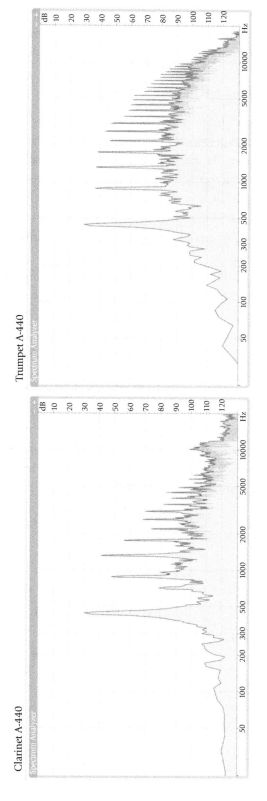

Clarinet A-440

Trumpet A-440

1760 Hz, and so on. To emulate these timbres with a synthesizer it would be necessary to reproduce each harmonic at the corresponding amplitude seen in the graph. Where things get tricky fast is when you analyze the spectrum over time and discover that, although these harmonics remain present, their intensities shift with changes in the player's performance techniques and in the natural decay of the sound.

Harmonics, Partials, and Overtones

Throughout this book we intersperse the terms harmonics, overtones, and partials, which might create the false impression that they are synonymous. Although similar, they each have specific definitions. All harmonics above the fundamental pitch, for example, are overtones but not all overtones are harmonics (see Figure 2.4).

Harmonics are sine waves at frequencies that make up the harmonic series and begin with the fundamental as #1 and continue with its integer multiples. All frequencies above the fundamental are considered overtones, but only those at integer multiples can be referred to as harmonics.

Partials is a catchall term for any frequency within a complex sound: that includes the fundamental itself plus any integer or noninteger multiple of it. If a frequency is produced below the fundamental, it is referred to as a *subharmonic*.

Unpitched instruments produce mostly noise—a complex wave with lots of overtones but few if any harmonics. Percussion is a good example of a sound comprising mostly inharmonic overtones with only a weak pitch center (ever tried to match the pitch of a snare drum with your voice?). That being said, most instruments produce at least some noise, even if just at the attack portion. For example, the initial attack of a piano is very percussive and noise-like before quickly transitioning to a decay with a strong pitch center (see Figure 2.5). Wind instruments, on the other hand, tend to have a bit of noise from reed or lip vibration and the breath passing through. Looking back at Figure 2.4 we see the combination of harmonic and inharmonic content in the spectrum.

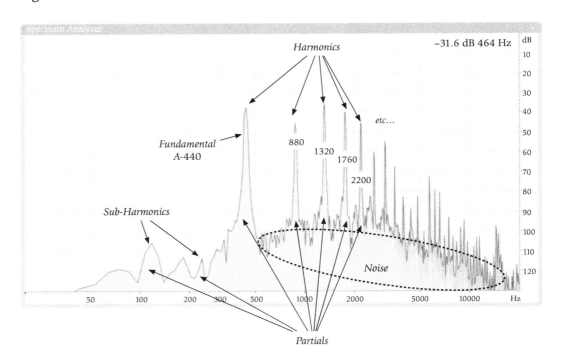

Figure 2.4 Labeling the components of a pitched complex wave.

Figure 2.5

Complex wave with a strong pitch center (left) versus one that is largely inharmonic (right).

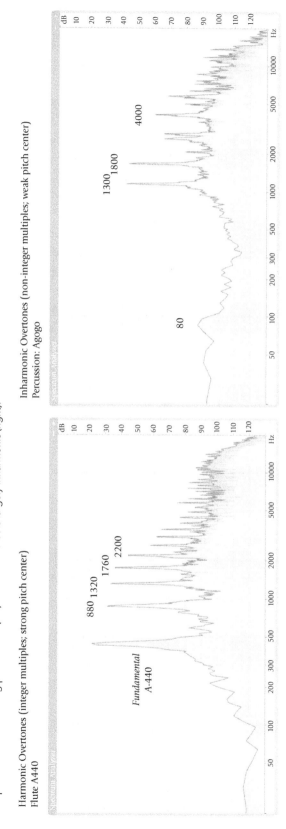

Harmonic Overtones (integer multiples; strong pitch center)
Flute A440

Inharmonic Overtones (non-integer multiples; weak pitch center)
Percussion: Agogo

The way all of these partials are created depends on how a particular instrument category produces vibrations. On a stringed instrument, the vibration of the string follows a predictable pattern. When viewed in slow motion, a string is seen to first displace fully in the middle (which produces the fundamental), followed quickly by a "node" forming in the middle, which acts like a dividing line of sorts with only the top and bottom halves of the string now displacing, the farthest points of which are called antinodes. Then two nodes with three antinodes form; then three; and so on. Each step increases the number of nodes and the ratio between the resulting wave and the fundamental frequency by one. The image on the left-hand side of Figure 2.6 shows this vibration pattern. The right-hand side shows the spectrum analysis of string vibration with several of the first of these same harmonics clearly defined.

Natural Harmonic Series in Music

All pitched instruments will produce harmonics in the same pattern of ratios relative to the fundamental; only the intensities of the harmonics will vary, giving each their characteristic timbre. So far we have only considered these harmonics in frequency values. When viewed on a musical scale (as in Figure 2.7) the harmonics reveal something interesting: their role as basis for musical scales and common chord progressions. This harmonic series is built off a low C2, and visible in the first four harmonics is the bedrock chordal progression of western tonal music: I–V–I (C–G–C). The equal tempered scale—the scale with which we are most familiar—is largely based on the natural harmonic series, but with twelve equal divisions of an octave rather than the naturally occurring integer multiples of the fundamental.

Complex Waveforms

All sounds in nature and from acoustic instruments have waveforms that are considered complex. By this it is meant that they are a compound of harmonics caused by vibrating and resonating components. In Figure 2.8 we see several common instruments and choral sounds dissected into their constituent harmonics. Way back at the turn of the nineteenth century, Joseph Fourier first recognized that complex sounds are really just an amalgam of partials, each of which is itself a sine wave. Therefore the sine wave (shown earlier in Figure 2.3) is effectively the basic building block of all sound, and when we view the dominant harmonics in Figure 2.9, we are looking at individual sine waves working together to create the timbre.

Phase

Phase describes the combination of two identical (or very similar) sounds separated by a time interval. Delay may be the unintended product of analog or digital audio system latency, sound picked up by separated microphones, or introduced deliberately with a delay-based-effects processor. Figure 2.9 shows how the degrees of rotation in a circle can be conveniently used to define the stages of development in a single wave cycle.

Phase can be described in more useful detail as one wave being combined with another at a certain degree point. Saying two waves are "out of phase" is less useful than saying

Figure 2.6

String vibration patterns (left) and spectral analysis of a plucked note on a nylon string acoustic guitar (right).

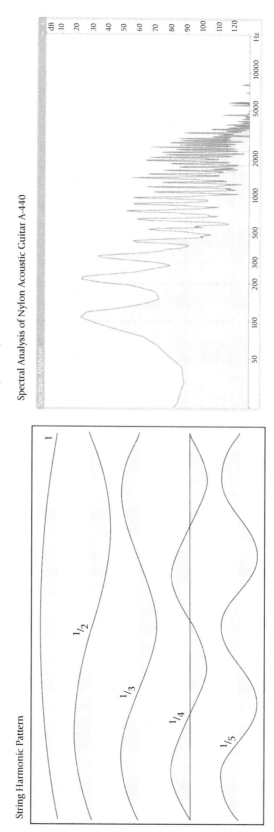

String Harmonic Pattern

Spectral Analysis of Nylon Acoustic Guitar A-440

Creating Sounds from Scratch

Figure 2.7
Harmonic series displayed on a musical staff.

Natural Harmonic Series

*note: Bb and F# are slightly "out of tune" relative to the commonly used tempered musical scale.

they are 90° or 180° out of phase, both of which have very different consequences (see Figure 2.10).

Remembering Fourier's lesson that a complex wave is just a whole lot of sine waves on top of each other at different frequencies, when two of the same complex waves are combined at a short delay, some of those sine waves will meet closer to 0° and result in constructive interference, meaning the sum of the two increases the amplitude at that frequency, whereas others will combine closer to 180° and result in destructive interference, a decrease in amplitude at that frequency. When viewed on a spectrum analyzer, spikes in amplitude occur at the points of constructive interference and dips occur at points of destructive interference. With some imagination the spectrum has the appearance of a hair comb with the tines pointed upward, so the effect has become known as *comb filtering* (see Figure 2.11).

In synthesis, comb filtering can be used creatively to color a sound, particularly when the amount of delay applied is changed over time in either a repeating pattern (perhaps defined by a LFO or an envelope shape). Listen to the demonstration in Audio Example 2.1.

In the preceding example, comb filtering is created simply by misaligning two complex waves when they are summed. A similar and more common effect in synthesis is slightly detuning an oscillator relative to another with both set to the same waveform. The change in wavelength that is due to the pitch shift creates a comb filter. Listen to the demonstration in Audio Example 2.2.

The spectrographs in Figure 2.12 reveal distinct differences in the amplitude levels of the harmonic components as a result of the detuning. In particular, note the (a) presence of a subharmonic below the 87-Hz fundamental that wasn't part of the original sound, the (b) attenuation of the fundamental, and the (c) increase in amplitude of the second harmonic. The before and after versions of this along with a video of the real time measurements are available at Audio Example 2.3.

Pitch

Pitch is a function of the frequency of a tone that is perceived when a sound is at least 2 ms in duration (this varies a bit with frequency, but is good enough for this conversation!). Shorter durations begin to sound more like noise and lose a strong pitch center.

Harmonics also play a role: We can hear a sine wave as having pitch, but it is much easier for us to identify when that sine wave is a fundamental reinforced by harmonics. Our ears are very good at discerning pitch: With experience, some musicians are able to hear the difference between a pitch that is "in tune" from one that is only a few *cents* sharp or flat. The term cents is used to describe the difference between two pitches separated by less than a semitone. For example, a note that is a halfway between a C and a C# is said to be 50 cents sharp of C (or 50 cents flat of C#).

Figure 2.8
Spectral analyses of familiar musical sources.

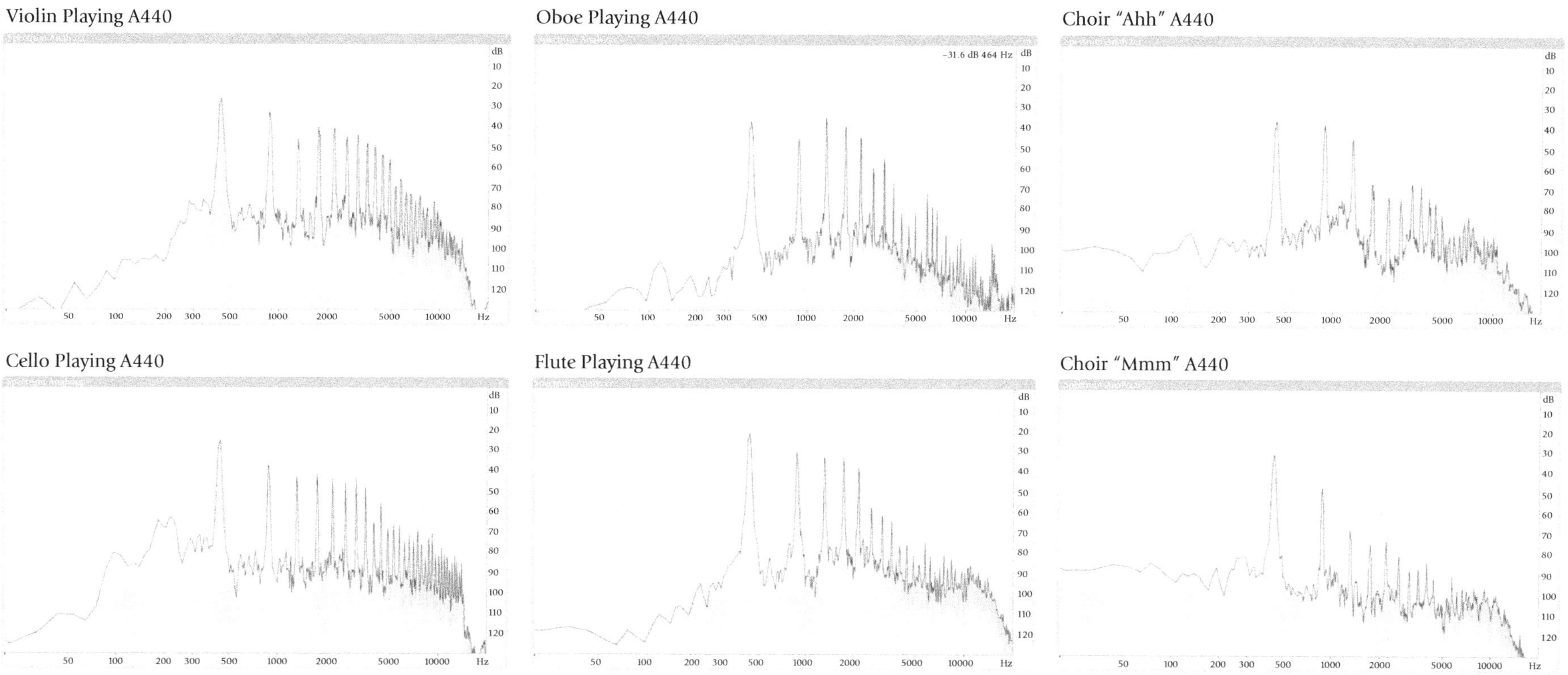

Figure 2.9
Degrees of a wave.

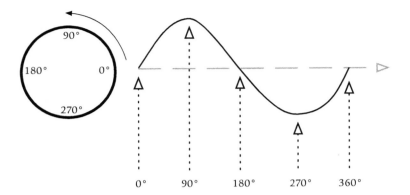

Figure 2.10
Summing sine waves at 0°, 180°, and 90°. When combined at 0° there is a doubling of amplitude, and thus a doubling of intensity. When combined at 180° there is a complete cancellation, thus no output. It is more common, however, that sine waves will meet slightly out of phase, somewhere other than 0° or 180°. At 90° there is attenuation but not complete cancellation.

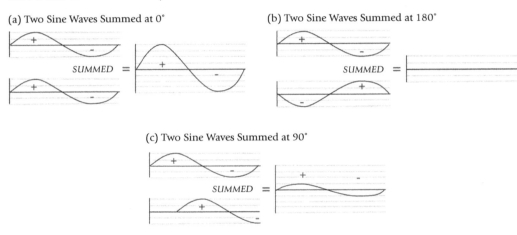

Figure 2.11
Comb filtering from two pink-noise tracks summed slightly out of phase as viewed on a spectrum analyzer.

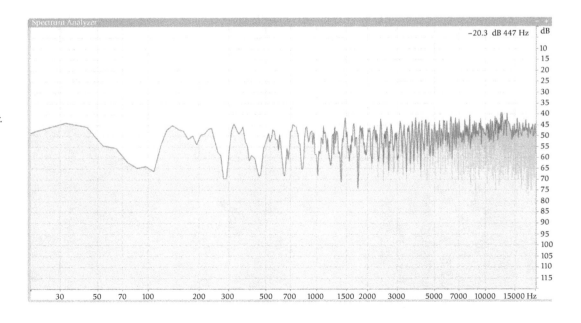

Understanding Sound and Hearing

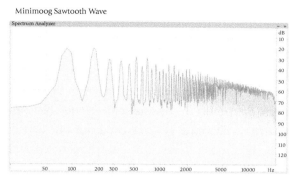

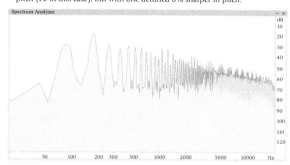

Figure 2.12
Original sawtooth wave (top left) combined with another sawtooth slightly detuned. Notice in the summed version the change in amplitude of the harmonics.

A creative use of the beating is a thickening of a sound by using two oscillators set to pitches separated by less than 50 cents, as seen above on the Arturia Minimoog. In this example Oscillator-2 is shifted up by about 2% and Oscillator-3 is shifted down by about 5%.

Doppler Shift

Although the name Doppler might first bring to mind a tool for meteorologists, the same principle is at play for an old technique that is especially near to the hearts of Hammond organ players—at least those who also own a Leslie cabinet or an equivalent with a rotating horn (see Figure 2.13). The Doppler effect is heard often in daily life when an emergency vehicle with a siren blaring comes toward and then moves away from you. A stationary or slower-moving bystander will have the impression that the pitch emanating from the vehicle shifts from a higher frequency to a lower one just as it is passing by. As the sound source is approaching the bystander, the rate at which the wave cycles are passing is greater than the actual frequency of the sound it's producing, and therefore a higher frequency is

View from rear of cabinet
(Logic's Vintage B3 Instrument)

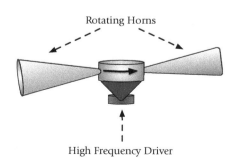

Figure 2.13
Leslie cabinet for Hammond organ with rotating horn for creating vibrato from Doppler effect.

heard. When the vehicle is moving away from the bystander the opposite occurs; the wave cycles are approaching slower, resulting in the perception of a lower frequency. Of course, anyone sitting in the moving vehicle will not perceive any shift because they are traveling at a velocity equal to the sound source and hearing the engine or siren at its actual pitch.

Figure 2.13 shows how the rotating horns of the Leslie coupled with the high-frequency driver (aka "tweeter"). From a switch on the organ console, a motor begins spinning the dual horns and creating a complementary Doppler shift as the horn on the left comes toward the listener and the horn on the right moves away. The low-frequency driver (aka "woofer") at the bottom fires down into a spinning drum (seen with an Electro-Voice RE20 mic at the bottom of the cabinet) shaped to create the Doppler shift in the bass frequencies. The rotary speaker effect has been modeled quite well in software like Logic and Native Instruments' (NI) Vintage Organs, although purists might argue to the contrary!

Amplitude

Zooming away from the amplitude-over-time plots so that individual wave cycles are no longer visible reveals the longer-term trend of a sound. This is as important to the character of the sound as the raw timbre itself. From the onset of a sound through its release, the intensity of any acoustic sound has a characteristic shape. Figure 2.14a shows how the amplitude of a plucked double bass evolves over time in two stages, the initial attack and a gradual decay.

With either a bowed string instrument or a wind instrument, two more stages are possible. Figure 2.14b shows four stages to a held note on a trombone: initial *attack* followed by the *decay*, a *sustain* stage followed by the note's *release*. In synthesis we can control these stages using a volume envelope with adjustments for the *duration* of the attack (A), the *duration* of the decay (D), the *level* of the sustain (S), and the *duration* of the release (R). More complex envelopes may also include *hold* (H) stages in one or more of the in-between points. A simple envelope will be labeled as ADSR; one that is more complex could be AHDSHR. (The volume envelope is covered in more detail in the next chapter.)

The volume envelope is as important as a sound's timbre for recognition if the sound design goal is emulating an acoustic instrument. Conversely, it can be exploited to creatively obscure the origin of a recorded sample. To demonstrate, let's consider what happens when the volume envelope is altered in an unnatural manner using a sample of a familiar acoustic instrument (Audio Example 2.4).

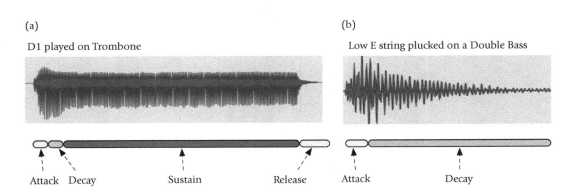

Figure 2.14 Volume envelope.

Piano and guitar both have percussive qualities, meaning they have an abrupt attack followed by a decay/release—no real sustain to speak of. When the attack is anything but abrupt or a sustain is inserted, the character of the original instrument is present but becomes obscured, perhaps to the point of familiar but unrecognizable. This is a useful way of creating unique sounds from a recorded sample, and unlike a timbre that comes from an oscillator, it retains an organic quality from the inherent complexities of a "real" sound.

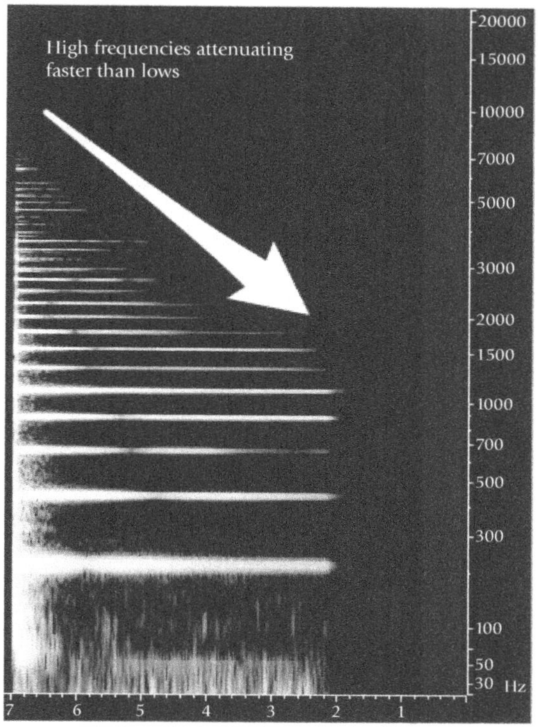

Figure 2.15 Spectral analysis of an A-220 played on piano and released after a few seconds of sustain. Note how the highest frequencies shown around 7000 Hz attenuate quickly, those below sustain longer until the dominant tones in the range of 220 through the fifth harmonic at 1100 (= 220 × 5) sustain for the full duration this note was played.

Timbral Shape Over Time

Envelope shaping is most commonly considered as a tool for amplitude control, but it can be used to great effect on any adjustable parameter on a synthesizer shaped over time such as (a) frequency-spectrum filtering, (b) intensity of an effect like chorus or flange, or (c) varying intensity of vibrato, just to name a few. (Audio Example 2.4 uses the same sample as that of Audio Example 2.3 but with an envelope controlling the shape of the LPF.) As evident in Figure 2.15, an acoustic sound's frequency content will change over time, so this control is useful in both simulating a real sound and designing something from scratch.

Human Hearing

Understanding the behavior of sound through the study of acoustics is needed to appreciate the physical aspects of the sound we create. As important, however, is the way in which our ears and hearing system collect and interpret sound. There are characteristics, we may even call them shortcomings, of our hearing system that offer means for some creative manipulation. Fortunately for those of us who understand what's going on, this kind of trickery continues to work even when you know it's happening! We have already touched on some, so let's look at others you might want to explore when creating sounds.

Phantom Image

Perhaps the strangest of these psychoacoustic phenomena results when the listener's head is facing two loudspeakers such that the head and two loudspeakers form an equilateral triangle. When two loudspeakers in this configuration reproduce the same sound, the brain receives signals from both ears simultaneously and assumes the source to be halfway between the loudspeakers where, curiously, no sound source exists—thus the term *phantom*. Amazingly, and fortunately for creative sound people, even knowing

Figure 2.16
Phantom image.

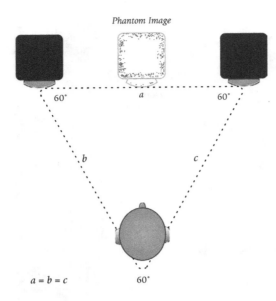

the actual location of the sound source does not diminish this effect (see Figure 2.16).

The location of the image is defined by both the loudness or intensity differences between the loudspeakers and differences in time of arrival. If the left-hand side is louder than the right within a range of 2–≤15 dB, the listener will perceive the sound source as being proportionally closer to the left-hand side; at ≥15 dB the sound will appear to be coming only from the left-hand side. The phantom will also be off center when the same sound is delayed in one channel within the range of roughly 0.2–1 ms.

Critical Bands

A simple yet effective way of adding depth and color to a sound is by exploiting another deficiency in our ears: the inability to detect two separate frequencies when they are close in frequency. Without getting too deep into the physiology of this effect, the cochlea—the primary organ of the inner ear—is where a sound is divided into its component frequencies and then sent to the brain for processing. Although there is roughly a single nerve cell for every audible frequency (more than 20,000!), these cells are grouped into sequential frequency bands across the spectrum, forming what are known as critical bands (see Figure 2.17).

When two frequencies fall within a critical band we do not hear them as separate pitches but rather as a vibrato-like oscillation between the two called *beating*. When two frequencies are just within the critical band, the oscillation is fast; as they get closer together, the oscillation slows. The rate of oscillation is equal to the difference between the two frequencies: 440 and 441 Hz will produce a beating of 1 Hz; 440 and 450 will produce a beating of 10 Hz (Audio Example 2.5).

As a creative application, timbres can be made to sound fuller by slightly detuning one oscillator against another when both are generating the same waveform (Audio Example 2.6). Why such a dramatic difference? The starting sounds are complex in nature, meaning they are made up of a fundamental and many harmonics. Not only is the fundamental being met within a critical band but so, too, are each of the harmonics, with different amounts

Figure 2.17
Division of critical bands in the human ear.

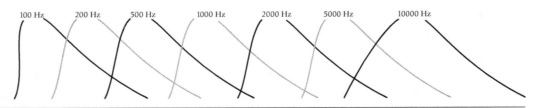

Approximation of Critical Band regions in human hearing

of beating at each position, resulting in a more rich and compelling timbre than summing oscillators at identical frequencies would achieve (see Figure 2.18).

Along with the creative implications, beating is particularly useful for assisting with tuning. When matching one string to another on a guitar, for example, the player will hold down a note on one string that is meant to be the same as an open string. Listening for the beating to slow down and eventually disappear makes matching two frequencies by ear simple.

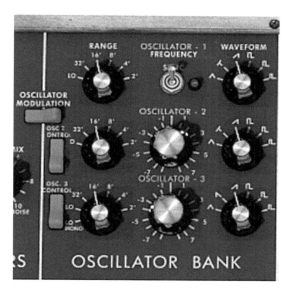

Figure 2.18
The Minimoog (screenshot of Arturia's Mini V) using three oscillators: Oscillator 1 is fixed "in-tune," Oscillator 2 is set slightly sharp, and Oscillator 3 is set slightly flat. A generic-sounding sawtooth becomes much richer through adjustment of only the pitch control.

Frequency Masking

When mixing timbres to create a patch or when crafting an overall mix of a piece of music, masking is arguably the most important concept to understand. Put simply, masking occurs when one sound conceals another. In the most obvious scenario, a louder sound will mask a quieter sound when they occupy the same frequency range. Masking can be avoided by distributing sound across the audible spectrum rather than having two or more sounds concentrated in the same frequency range. It is important when designing a patch to think about its role in the arrangement and choose starting pitches or use filtering to isolate it into a frequency range where it is needed. Listening to the sound both alone and in the context of the larger mix is how mix engineers work to fit everything into its appropriate spot. Being conscious of a timbre's role in the bigger picture while it's being designed makes for a more coherent mix, and an easier mix job when that time comes. Thinking about using the overall audible spectrum is also a great way to arrange a song or piece as you add elements.

A well-crafted mix and/or orchestration will spread elements with balanced levels across a wide frequency spectrum, thus minimizing masking effects (see Figure 2.19). It is tempting to add too much to what we call the "mud range" (100–300 Hz) because so many elements common to modern music (guitar, bass, keyboards, and male vocal, just to name a few) occupy that area if not carefully arranged in the composition or filtered to "sit" properly. A mix that spans the audible spectrum sounds full and tends to be more interesting to the listener.

Figure 2.19
Segments of the spectrum where your mix elements should be targeted to minimize problems related to masking.

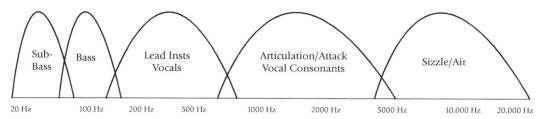

Figure 2.20
Shaping over time two sounds occupying the same frequency range to avoid masking.

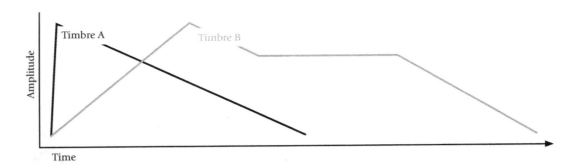

It's also important to note that louder elements of a mix mask a wider frequency range than do quieter elements. Those who are fans of mixing their bass heavy, take note! Low frequencies will mask higher frequencies as their level is increased, so be careful not to add more than you need, and avoid occupying the bass and sub-bass ranges with multiple sounds—that just makes for a mud puddle at the bottom. A common challenge for mix engineers working in rock is sorting out space for both the kick drum and the bass guitar. The decision comes down to which will truly act as the bass instrument. Curiously, however, when the frequency "budgeting" is complete, the listener's perception will still be that both the kick and the bass are sitting at the low end of the mix—another brain trick!

Masking may also be avoided when layering two timbres in the same frequency range by giving them very different volume envelope shapes (see Figure 2.20). Timbre "A" comes in with a fast attack and medium decay; timbre "B" has a much slower Attack (making A more audible) and extends well past timbre "A." In the bass/kick drum analogy, engineers may use a compressor on the bass channel side-chained to the kick so that its attack drops out to make way for the short kick and then recovers when the kick sound is gone. Careful crafting of this balance will create the illusion of both sounds being present and equal in intensity.

Loudness

In one sense, the concept of loudness is largely subjective: What is loud to you might not seem loud to me. Music or sound you dislike always seems louder when objectively measured at the same intensity as one you would describe as pleasant. The kind of loudness we are concerned about here, however, has to do with how harmonic content and the duration of a sound affect its perceived loudness.

Loudness Relative to Harmonic Content

Earlier it was stated that loudness and amplitude are related. That is true to an extent. We also hear a sound that is richer in harmonics as being louder that one that is closer to a pure tone, like a sine or triangle wave (see Figure 2.21). A lead instrument's presence will benefit from starting with a more complex wave, processing the sound in a way that adds resonance or adding a touch of overdrive or distortion (a process that adds odd harmonics to the frequencies present).

Figure 2.21

The pure tone on the left (a sine wave alone) will be perceived as softer in loudness compared with the same note with a patch of a more complex wave.

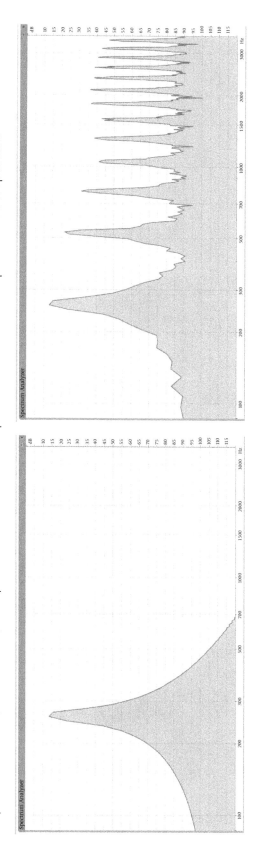

Creating Sounds from Scratch

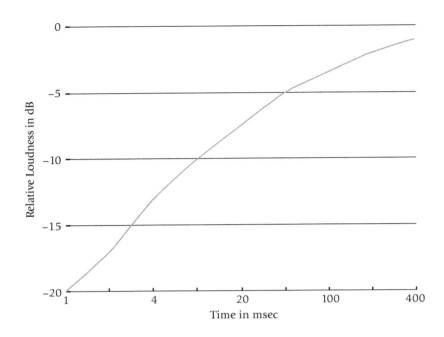

Figure 2.22
Loudness relative to duration, an important consideration for designing percussive sounds.

Loudness vs. Duration

Masking can also pertain to the duration of a sound, specifically how a sound's length affects its loudness. Figure 2.22 shows an estimation of the perceived loudness of a sound relative to its length in milliseconds. According to this graph, it is not until the sound reaches almost half a second (400 ms) that its actual intensity begins to correlate with how loud we perceive it to be. In other words, a very short sound intended as a percussion element that falls on the short end of this scale may be masked by louder or longer sounds also in the mix. Shaping the decay or release times or adding reverb may be enough to extend the duration and thus increase the sound's perceived loudness.

Loudness Relative to Frequency

It's also important to understand that we do not hear all frequencies equally at different overall volumes. A familiar example is all the extra low end you hear when you turn your stereo speakers or headphones up. The additional bass is not a product of your sound system; our ears are just better able to hear the long wavelengths of bass frequencies when the volume is turned up to a high level.

Figure 2.23 shows the Equal Loudness Contours, a very important chart used for everything from microphone design to developing codecs like MP3, and is useful—at least on an intuition level—when one is building up a mix. The curves on the graph represent an average of human hearing at volume levels spanning from the lowest curve at about 20 dB-spl to the highest at 90 dB-spl. For our purposes, there are a couple fundamental points to take away from this chart:

1. Our perception of the balance of low to high frequency is never even, but it is more so at higher volumes. Conversely, we struggle to hear bass frequencies at low listening levels. Be careful mixing at a quiet level something you intend to play back at 90–100+ dB-spl at a club; the bass will

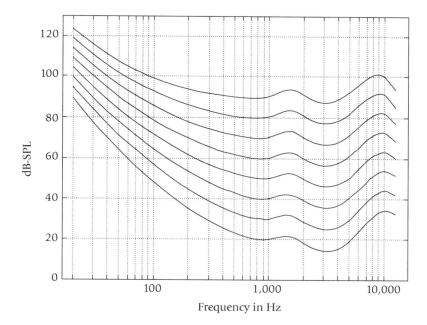

Figure 2.23
Equal loudness contours (ISO 226).

be out of control at the higher volume. And don't listen too loud in your studio or moderate listeners will hear too *little* bass in your mix. Peaking at 85–90 dB-spl when monitoring in the studio is a good target zone. If you cannot listen at that level all the time, be sure to check your balance in that range before finalizing a mix.

2. There are two zones in which we are particularly sensitive. 3–4 kHz is the most sensitive area of our hearing, it is where our ear canal naturally resonantes. Incidentally, and not surprisingly, this is where consonants, critical to our speech, reside. Adding anything here with equalization (EQ) or resonance on a filter will bring it to the forefront of a mix, but use it sparingly, like you would a strong spice. The next pronounced dip occurs around 12 kHz (the upper limit of the graph in Figure 2.23). This is a good spot to exploit because we are sensitive there but not nearly as much as 3–4 kHz (technically it is the third harmonic of the 4-kHz resonance).

Loudness Perception (RMS/Peak)

The perception of loudness is, of course, largely subjective. One person's "loud" may be another's "just right." In this context, however, we are more interested in the real difference between the loudness of one sound versus another. Generally, complex sounds that contain more harmonics (particularly in areas of increased sensitivity shown in Figure 2.23) are heard as louder than those that rely more on the fundamental.

All workstations have meters, and the standard mode is for showing peak values. Peak values are useful for seeing if a signal is going into distortion but it has little to do with loudness. Root-mean-square (RMS) modes are commonly available as options and, in general, are better indicators of loudness because they average the peak value over a

Figure 2.24
Peak (darker gray) and RMS (lighter gray) meters on Logic Level Meter plug-in.

period of time, and thus take into account the concept of loudness relative to duration that we already covered (see Figure 2.24). They do not, however, incorporate the equal loudness contours. Volume units (VUs) are becoming less common but are also based on an averaging over time.

Missing Fundamental

Our brain is really good at interpreting data received from our various senses, but sometimes it can be fooled. Phantom images are a good example: Put on a pair of headphones with a signal sent equally to both ears and our brain deceives us into thinking there is a sound right in the middle. Weird! But wonderful. Clearly an artifact of evolution that needed no tweaking: Fortunately for humankind, creatures that wanted to have us for dinner never figured out how to timecode-synchronize their roar from equal positions to our left and right so that we would be convinced the threat was right in front of us! Another example that has no obvious evolutionary advantage is the missing fundamental.

Our brain has learned the harmonic series so well that when something it deems important is missing, it adds it in! That's right, if it receives a signal for a harmonic series with $2x$, $3x$, $4x$, ..., and so on, it is convinced the fundamental ($1x$) must be there somewhere and drops it in like having your own personal cochlear Moog. This trick has been exploited for years on pipe organs where a stop might be labeled "32 ft" when it is clear, just by looking, that the largest pipe cannot be longer than 16 ft. Selecting "32 ft" will instead produce a low perfect fifth simulating harmonics two and three, leaving our brain to add in the octave below and round out the first three harmonics of the series. Pretty neat and something we can clearly exploit.

Effects like Waves' MaxxBass (Figure 2.25) use the concept of missing fundamental to make smaller monitor speakers create the illusion of reproducing frequencies below that of which they are physical capable. The process is the same: The plug-in looks for low frequencies, synthesizes a copy a fifth higher, and our brain adds in the lower octave. This will work on larger speakers as well but it's most impressive on small speakers, because they seem to be working at a lower frequency than is physically possible. Overuse will quickly add murkiness to the bottom of you mixes but, used sparingly, it can be a nice touch.

Understanding Sound and Hearing

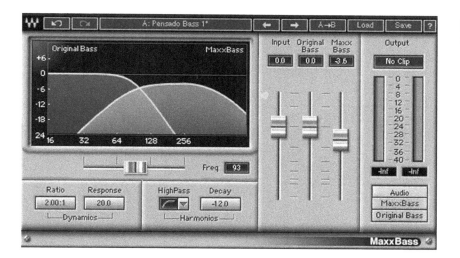

Figure 2.25
Waves MaxxBass.

Conclusion

So that's it for our coverage of physical and psychological aspects of sound. We have barely scratched the surface of these deep and interesting topics. Hopefully this has provided you some insight that will help guide you in your sound design and mixing.

Helpful Facts

Inharmonic vs. Enharmonic—These two terms are very different, which is unfortunate given that they sound identical if not well enunciated. *Inharmonic* refers to overtones that are not integer multiples of the fundamental. A♯ and B♭ are *enharmonics*; meaning they are spelled differently but correspond to the same pitch on an instrument, the same black key on the piano keyboard.

Overtones are by definition *over* the fundamental. Therefore, the first overtone of A-440 is 2 × 440 or 880 Hz. The first *harmonic* is 1 × 440 or 440 Hz—thus the first harmonic is also the fundamental. A common point of confusion—and for good reason—is that the first overtone is also the second harmonic!

3

The Tools of the Trade

Introduction

All synthesis techniques we explore in the following chapters contain a similar structure: (1) some means by which sound in generated or played back, and (2) some means by which that sound is modified. This chapter serves as an overview of sound design processes: the sound generators, the modifiers, and the hardware and software used in the process. Each chapter that follows goes into greater depth on the most common forms of synthesis in use today.

Sound Generators

There are essentially two primary sound sources in all of music synthesis: the oscillator (which forms the foundation of subtractive, frequency modulation and additive synthesis) and digital playback (the basis of sampling, physical modeling, wavetable, and granular.) The ways in which they are applied to create the finished sound varies dramatically. Table 3.1 lists the forms of synthesis covered in this book followed by the generator type and a description of the generator's product.

Oscillator (VCO)

The traditional analog method of generating sound electronically is with an oscillator, an electronic component that produces an electrical alternating current (ac). A voltage-controlled oscillator (VCO)—the type used for music synthesis—accepts a direct current (dc) as control information to vary the number of cycles generated per second (i.e., changing the pitch). This dc signal is referred to as the control voltage and is labeled as "CV." Instruments made by Moog, Sequential Circuits, Roland, Oberheim, and ARP followed a convention of each octave being represented by a change of ±1 V dc. In other words, if a VCO reproduces an A2 at 110 Hz when fed with 2 V dc, then it will play an A3 at 220 Hz with 3 V dc. Keyboard instruments following this convention will transmit a CV as ±1/12 V dc steps to represent the familiar twelve semitone-per-octave black and white keys that make up an octave.

Creating Sounds from Scratch

Table 3.1 Generators and Products for Popular Forms of Synthesis

Type	Generator	Product
Subtractive	Oscillator(s)	Triangle, pulse/square, and sawtooth waves, and noise
FM	Oscillator(s)	Modulated sine waves, "operators"
Additive	Oscillator(s)	Summed sine waves
Sampling	Sample playback	Realistic reproductions of recorded sounds
Physical Modeling	Mathematical algorithms	Complex waves with user control of harmonic structure elements
Wavetable	Resynthesized samples	Waveforms based on an analyzed sample
Granular	Samples divided into small fragments called "grains"	Small grains with flexible pitch and tempo manipulation reproduced in a user-controlled order.

The basic waveforms generated by a standard oscillator consist of sine wave, triangle wave, square (or pulse) wave, and sawtooth wave as well as white (or pink) noise (see Figure 3.1).

Figure 3.1
Standard oscillator waveforms and their harmonic content (reference Audio Examples to hear how they sound).

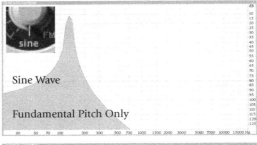

Sine Wave

Fundamental Pitch Only

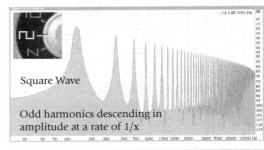

Square Wave

Odd harmonics descending in amplitude at a rate of $1/x$

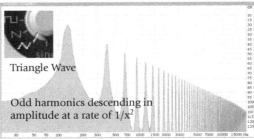

Triangle Wave

Odd harmonics descending in amplitude at a rate of $1/x^2$

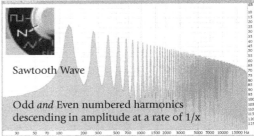

Sawtooth Wave

Odd *and* Even numbered harmonics descending in amplitude at a rate of $1/x$

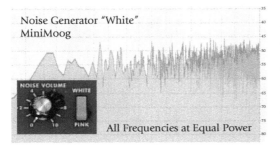

Noise Generator "White"
MiniMoog

All Frequencies at Equal Power

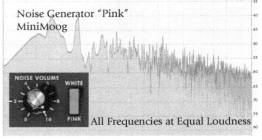

Noise Generator "Pink"
MiniMoog

All Frequencies at Equal Loudness

Sine Wave

The sine wave is the most basic waveform, consisting of only its fundamental pitch. It represents simple harmonic motion, a smooth fluctuation of compression and rarefaction with no sudden disruption to the flow (see Figure 3.2). As a building block it might be useful to fill in a lower octave of a more complex sound but otherwise, on its own, its application as a sound is limited. It is very helpful, however, as a modifier shape and, as you will see, it is the basic building block of *all* sound.

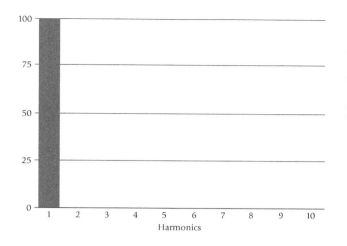

Figure 3.2
Sine-wave fundamental only; no harmonics (Y axis = amplitude percentage, X axis = harmonics of the fundamental).

Forming what is known as a Fourier series, complex waveforms—basically any waveform that is *not* a sine wave—can be deconstructed into many individual sine waves at varying intensities and frequencies. A pitched sound consists of a sequence of harmonics beginning with the fundamental that defines its pitch. The triangle, square, and sawtooth waves' spectra depicted in Figure 3.1 reveal their harmonics as spikes in amplitude at frequencies made up of integer multiples (1x, 2x, 3x, 4x, etc.) of the fundamental—individually these harmonics are sine waves. The noise examples also contain sine waves but with no repeatable pattern or discernible fundamental (see Figure 3.2).

As we will see in the next chapter, the process of subtractive synthesis involves taking a complex wave (one rich in harmonics) and carving out unwanted elements until the desired timbre is achieved. Because the sine wave alone contains only a fundamental pitch it is not uncommon for subtractive instruments to leave the sine wave off their list of VCO waveforms in favor of those that are inherently more colorful.

The sine wave really shines in FM (covered in depth in Chapter 5), where surprisingly complex timbres develop from simply modulating one with another—in other words, as few as two sine waves are needed to create spectrally diverse timbres. Additive synthesis (covered in depth in Chapter 6) also makes extensive use of the sine wave as a building block of more complex sounds, effectively a means for constructing a Fourier series to your taste.

The triangle wave is mostly a fundamental frequency with a slight presence of odd harmonics decreasing in intensity at an exponential rate of roughly $1/x^2$ (with x being the number of the harmonic in the harmonic series) (see Figure 3.3). On its own it can

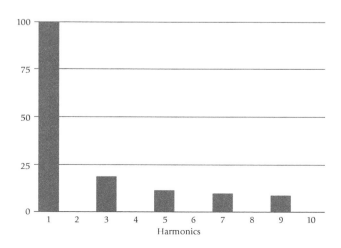

Figure 3.3
Triangle-wave odd harmonics descending in amplitude at a rate of $1/x^2$ (Y axis = amplitude percentage, X axis = harmonics of the fundamental).

Figure 3.4
Square-wave odd harmonics decreasing in amplitude at a rate of 1/x (Y axis = amplitude percentage, X axis = harmonics of the fundamental).

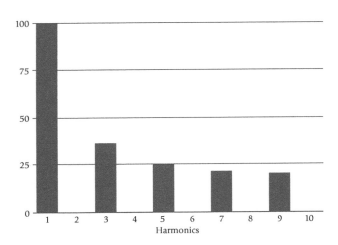

be an option as a timbral building block, but given its weak overtones it is a bit less useful than pulse or sawtooth waves for subtractive synthesis. It is, however, like the sine wave, a standard option on a LFO and is offered as the cleanest option on instruments like the Minimoog.

A square wave has equally sized positive and negative stages (symmetrical) in each wave cycle because of its composition of only odd harmonics that decrease less abruptly than those of the triangle wave at a rate of 1/x (see Figure 3.4). A pulse wave is best described as an asymmetrical square wave, one that has a transition point earlier than the halfway point so that the positive stage is shorter than the negative stage. The term *duty cycle* can be used to describe the position of the shift: 50% is a square wave, 10%–40% is a pulse wave. The shift away from the symmetrical square wave adds variation to the harmonic content, most notably as a *comb filter* in the higher harmonics, as seen in Figure 3.5.

A pulse-wave shape is colorful on its own but can really stand out when the duty cycle percentage is shifted over time by an envelope generator or LFO using what is called *pulse-width modulation*.

It is clear from the square- and pulse-wave spectrograph images of Figures 3.5 and 3.6 that the dominant harmonics are indeed the odd-numbered (F × 1, × 3, × 5, etc.) ones. The attenuation or omission of every other harmonic produces what might be described as a hollow quality, not unlike that produced naturally by a cylindrical bore-shaped instrument like the clarinet. In fact, a clarinet-like sound is relatively simple to approximate this way (the basis of Sound Design 4.1 in the next chapter).

Sawtooth Wave

The most complex of the standard waveforms is the sawtooth wave. Its dense collection of harmonics makes it a popular starting place for subtractive synthesis (detailed in Chapter 4). Whereas the triangle wave includes all integer-multiple harmonics (F × 1, × 2, × 3, etc.) descending exponentially to intensities far lower than the fundamental, and the square wave contains only odd harmonics descending linearly in intensity, the sawtooth wave includes *all* harmonics decreasing in intensity at a rate of roughly only 1/x (see Figure 3.7), and thus they become a prominent characteristic of its timbre.

The classic 1980s synth strings or synth brass patches were typically constructed with one or more sawtooth waves mixed together, often with each oscillator detuned slightly from one another to create a thicker sound. In Chapter 4 you will find Sound Design 4.2 that describes the creation of an '80s "buzzy" synth by sawtooth waves in a subtractive model using the Prophet V synth.

Square Wave

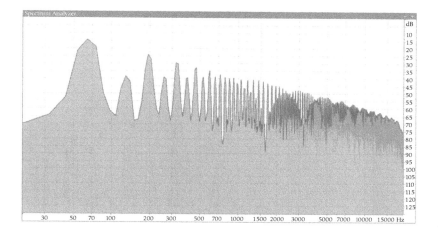

Wide Rectangle Pulse Wave

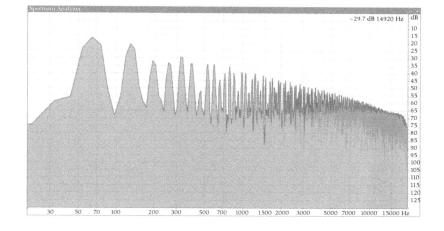

Narrow Rectangle Pulse Wave

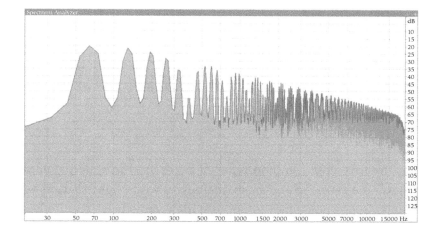

Figure 3.5
Comparison of square and pulse waves—notice how the odd harmonics (1, 3, 5, etc.) of the square wave are more dominant than the even harmonics. Holding a straightedge vertically across all three spectra images reveals the differences in intensities at each harmonic, thus accounting for the timbral differences.

Noise

It may seem strange to list noise as a desirable component, but it is an essential ingredient (1) when creating percussive sounds, (2) when accentuating the attack portion of a patch, or (3) when filtering the noise itself for use in sound design as a special effect (the sound of wind is a classic example).

Creating Sounds from Scratch

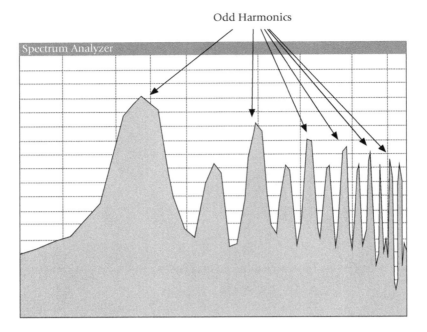

Figure 3.6
Odd harmonics dominant in square wave.

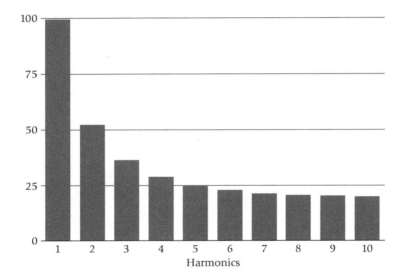

Figure 3.7
Sawtooth-wave odd and even harmonics descending in amplitude at a rate of 1/x.

The two most common types of noise used in sound applications are "white" and "pink." The term "white" is an analogy to white light that, like the noise, contains all frequencies at equal power. Because high frequencies require less power to be heard at a loudness equal to low frequencies, they sound far more prominent. Musical octaves are a doubling of frequency (20 to 40 Hz, 40 to 80 Hz, ... 1000 to 2000 Hz). Pink noise is made so that, starting at the lowest octave of frequencies (20 to 40 Hz), each successive octave uses half the power because its content is twice as large. The result sounds harmonically neutral or "flat," with 20 Hz (or at least as low as your monitors can reproduce!) sounding equal in level to 20 kHz—*if* your ears are still young enough to hear that!

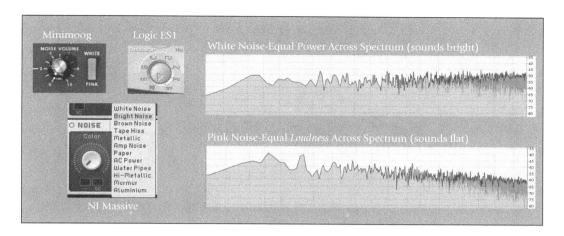

Figure 3.8
Noise sources: comparing white and pink noise.

Most instruments offer just white noise; others, like the Minimoog, have a switch for white or pink (see Figure 3.8); modern instruments, like Native Instruments' Massive, have a more extensive list of options.

An example of using noise in a patch is when simulating the sound of a synth snare drum. A waveform like a triangle can be used to represent the pitched resonance of the air inside the drum and the vibration of the shell, and noise can be used to represent the stick-hit attack and the rattle of the snares, the unpitched elements of the drum's timbre.

Digital Playback

Wavetable Synthesis

With the advent of digital synthesis came options beyond analog oscillators, and the sound palette grew exponentially. Stable oscillators outputting pure sine waves offered modulation of one by another, bringing about the timbral complexity with the elegantly simple components of FM synthesis (covered in Chapter 5). Sine waves were the only building blocks necessary to generate sound in FM, so the demands on the digital processor were negligible. With digital storage capacities increasing and costs coming down, the stage was set for storing and reproducing far more complex timbres, ones that strived to make synthesizers produce sounds reminiscent of or strikingly similar to acoustic instruments.

The first step in this direction was wavetable synthesis (covered in detail in Chapter 9) in which single-cycle complex waves were stored and looped under the shape of a volume envelope. Demands on storage and processing were very small, making this a perfect solution for the time. An early example of this technology was the Prophet VS (see Figure 3.9). As a means for shifting the wave over time, Sequential Circuits introduced vector synthesis (hence the "VS" in its name), a method by which the player could select four different waves from a wavetable list and place them as four regions around a joystick. The player could simply move the joystick around to shift the balance between the different timbres.

Figure 3.9
Wavetable synthesis on Sequential Circuits' Prophet VS—three waves shown from the wavetable of Arturia's plug-in version of the instrument (Prophet VS photo courtesy of Sequential Circuits).

  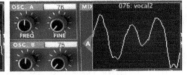

Sample Replay/Sampling

Moore's Law—named for the cofounder of the Intel Corporation—foresaw a trend as far back as the 1960s that computer processing power and memory storage would increase in the future on an exponential scale. So far his prediction has been accurate (although experts predict a slowdown to begin within the next decade following this book's publishing). This trend has benefited musicians too, of course, and on the heels of a wavetable's ability to store and reproduce single-cycle wave files, sampling came along for storing and reproducing recordings that were made up of far more than a single cycle, greatly improving the realism of sounds a synthesizer could produce. Initially this meant reducing sound files to low bit and sample rates and to samples far shorter than modern standards, but Moore's Law would see to it that those limitations would diminish with advancing technology. Sampling synthesis became a highly effective form of sound design and is covered in detail in Chapter 7.

Digital synthesis continued to evolve to satisfy sound designers restricted by the limited options for manipulating recorded samples. Additive synthesis (Chapter 6) was a technique that involved assembling a sound's harmonic structure, essentially Fourier analysis in reverse. Physical modeling (Chapter 8) was yet another approach that employed sophisticated mathematical models to describe the characteristics of a sound's generator (like hammers connected to the keys of a piano) and its corresponding resonator (the piano's strings).

Sound Modifiers

Whether the starting sound is generated with a VCO, sample replay, or wavetable, there are many tools common among instruments to further sculpt its timbre and guide its shape over time. Such tools are referred to as *modifiers*. If you consider that sound in its most basic form is a collection of waves at different frequencies and intensities varying over time, it is logical that the categories of modifiers will be those that affect the

Table 3.2 Modifiers: Common Sources and Destinations

	LFO	EG	VCF/LPF	Delay Effects
Pitch	•	•		
Amplitude	•	•		
Timbre			•	
Depth				•
Shape over time	(to VCF or VCA)	(to VCF or VCA)		

frequency content, those that affect amplitude, and those that shape either or both over time. Let's take them in that order.

Frequency-Based Modifiers

Low-Pass Filter

In Sound Design 4.1 (Chapter 4) we will begin with a square wave to approximate a square wave. On its own it doesn't sound much like a clarinet, in part because of the intensity of the high frequencies. To help get us closer we used the ubiquitous modifier: the *low-pass filter* (LPF). Although a confusing name at first glance, its meaning is quite literal: a filter that allows low frequencies to *pass* through while blocking high frequencies that exceed a specified point, a location referred to as the *cutoff frequency*.

The LPF is of utmost importance to the circuit design of analog synthesizers in that its characteristics are largely what gives a particular instrument its unique sound. Moog, ARP, and Oberheim synthesizers (just to name a few) all were distinguishable by and revered for their unique filter qualities. Digital synthesizers therefore often allow the user to select from a variety of filter options to further customize this effect or choose a design that emulates a classic.

In this screenshot of Logic's ES1 (Figure 3.10) the filter is currently set to 24 dB/octave "fat," which is said to emulate the Oberheim filter. One step counterclockwise shows a

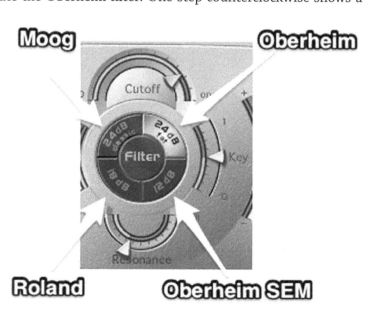

Figure 3.10
Filter options in Logic's ES1 synthesizer.

Creating Sounds from Scratch

Audio Example 3.1
Explore the creative possibilities of a filter with Logic's ES1 synthesizer as an example.

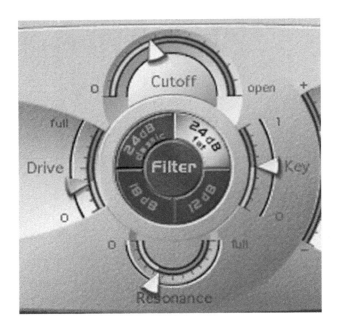

24-dB "classic," which is modeled after the Moog filter. Below are lower-order 12- and 18-dB filters emulating the Oberheim SEM and Roland TB-303, respectively.

Why the differences in filter sound among manufacturers? An analog LPF requires the audio signal to be combined with itself at a slight delay through the use of a resistor–capacitor (RC) circuit that introduces a phase shift that increases the amount of attenuation as the frequency increases. The untouched area is known as the *passband* and the area being attenuated is called the *stopband*.

Modifying the circuitry enables the steepness of the filter's slope to shift in increments of 6 dB: 6 dB/octave is "first order," 12 dB/octave is "second order," 18 dB/octave is "third order," and 24 dB/octave is "fourth order" (see Figure 3.11).

The introduction of a phase shift into the audio signal produces a distortion of the original wave in the transition area where the passband and stopband intersect. As the "order" increases, so does the intensity of this distortion. The contrasting circuit designs alluded to earlier generate different artifacts in this region that can be quite musical.

Figure 3.11
Filter cutoff orders: first, second, third, and fourth = 6, 12, 18 and 24 db/octave.

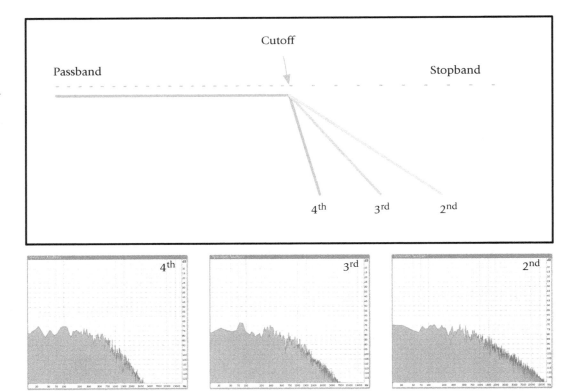

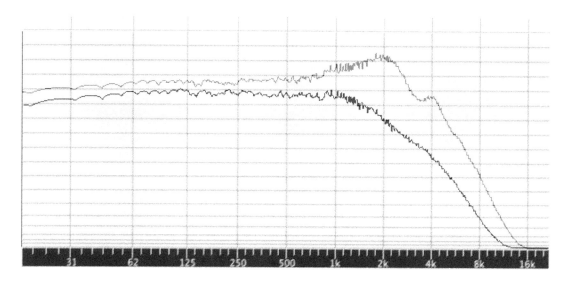

Figure 3.12
Pink noise with LPF no resonance and 100% resonance (Sound Source: Logic's ES1 synthesizer).

If this were the end of the LPF story, analog synthesis would be pretty boring. The behavior of the different analog filter designs is fairly consistent when simply attenuating high frequencies (see Figure 3.12). Where things start to get very interesting is when you turn the *resonance* control (sometimes called *emphasis*), and the true character of designs by Robert Moog, Alan R. Perlman (ARP), Tom Oberheim, and others, becomes evident.

Turning up the resonance control increases the intensity of frequencies around the crossover point (see Figure 3.13. At low levels this can add some bite and color to the sound; at higher levels the crossover area is far more prominent than the passband region resulting in a dramatic shift in timbre—when pushed far enough, the filter resonance

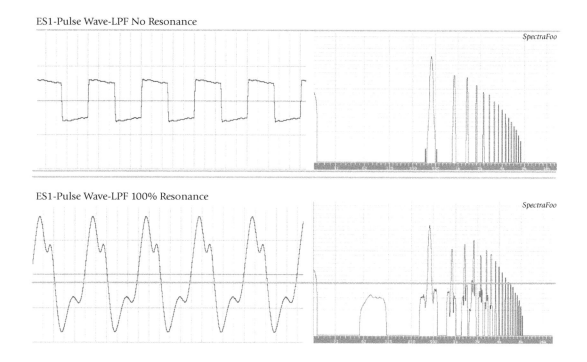

Figure 3.13
Effect of resonance on LPF using Pulse/Square Wave (Sound Source: Logic's ES1).

may even begin to self-oscillate! A curious side effect here is that the filter may self-oscillate even if you turn off the VCOs feeding into it! Some great sounds are possible this way. Let's look at how to use resonance in Sound Design 3.1 by using a free software instrument, TAL NoiseMaker.

Sound Design 3.1—Exploring Resonance and Self-Oscillation Using Pink Noise

High-Pass Filter

The high-pass filter (HPF) is essentially the opposite of the LPF in that it allows the high frequencies to "pass" through while blocking the low frequencies (see Figure 3.14). This is particularly useful for sounds that you want to emanate from the background of your aural soundscape or appear as a light texture. When you are layering sound, the HPF can be useful in segregating a timbre into a segment of the spectrum that spans from the cut-off frequency of the filter up through the highest overtones of the patch. The resonance characteristics on the HPF are different than those of the LPF but equally as important as a creative element.

As with the LPF, interesting results can be achieved by modulating the HPF with an envelope or a LFO. Many synthesizers do not include a HPF, but they are readily available in any digital audio workstation or digital mixers and many analog mixers as well.

Bandpass Filter

A *bandpass filter* (BPF) can be of value when your goal is to limit the sound to a portion of the frequency spectrum—again, as with the HPF, useful for segregating a particular timbre or patch into a restricted range, in this case one that sits in middle of the frequency spectrum, not reaching the extreme high or low areas (see Figure 3.15). You can think of a BPF as a combined HPF and LPF. Resonance control, as expected, will bring out more colors. It might be useful to blend a BPF with a version of the timbre with a broader frequency range. Some instruments, like Logic's ES2, allow blending between two filters. A similar effect can also be accomplished by applying the BPF to one oscillator while leaving another untouched.

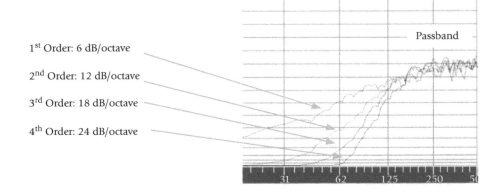

Figure 3.14
HPF orders from Spectrograph.

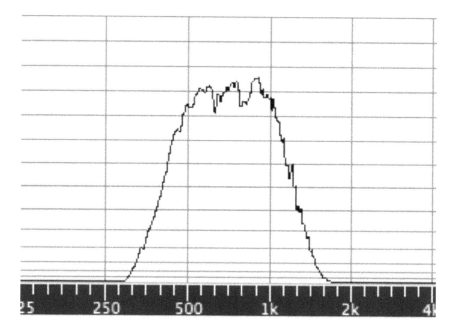

Figure 3.15
BPF 500–1000 Hz.

Notch Filter

Effectively the opposite of the BPF is a *notch filter*, although it typically works in a narrower range (see Figure 3.16). The notch filter may be used as a corrective tool, perhaps minimizing a frequency that is interfering with your desired sound, or as a creative tool, opening space for another sound element to fit through. Adjusting the resonance control will alter its bandwidth. As like other filter types, modulating the frequency location with envelopes and LFOs can yield dynamic interest.

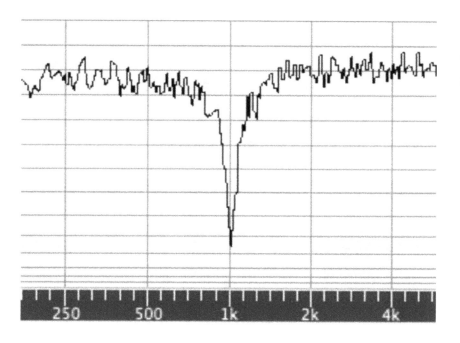

Figure 3.16
Notch filter centered at 1 kHz, narrow bandwidth.

Creating Sounds from Scratch

> As mixing tools, the LPF, HPF, and BPF can be useful to segregate a patch into the layer where it is needed. Figure 3.17 illustrates a hypothetical mix of instruments and their regions of the frequency spectrum.
>
> **Figure 3.17**
> Using filters to segregate a mix.
>
>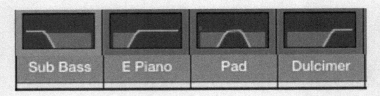

Shelving Filters

More commonly used as a mixing tool, the shelving filter's name is taken from the visual shape on the frequency spectrum (see Figure 3.18). It is used to boost or attenuate a range of frequencies above or below a given point. A high-frequency shelf boost can be helpful for adding presence to a sound by bringing out its overtones. A low-frequency shelf boost can bring out warmth in a sound that is too thin, or attenuated to function similarly to a gentle HPF.

Allpass Filter

Seems like a tool without a job, doesn't it? Filters are meant to block something, so what is the point? On its own it is not all that impressive. The *allpass* filter lets all frequencies through but introduces a delay that causes a phase shift when recombined with the original signal—because low frequencies have longer wavelengths, the effect will be more dramatic on the smaller high frequencies. Resonant peaks are a defining characteristic of allpass filters, the number of which is tied to the design of the filter: two-pole, four-pole, and eight-pole will produce two, four, and eight resonant peaks, respectively. Lining up several allpass filters in series will produce dramatic results. The *phaser* effect, as an example, depends on the allpass filter.

Not every instrument will offer all of these filter types. See Figure 3.19 for the filters available on four popular software synthersizers.

Figure 3.18
Shelving filters in Synthmaster 2.6.

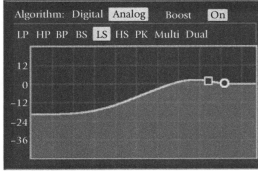

LS-Low Shelf (Analog Emulation)

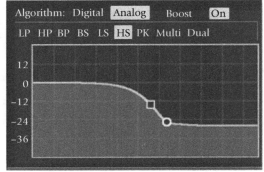

HS-High Shelf (Analog Emulation)

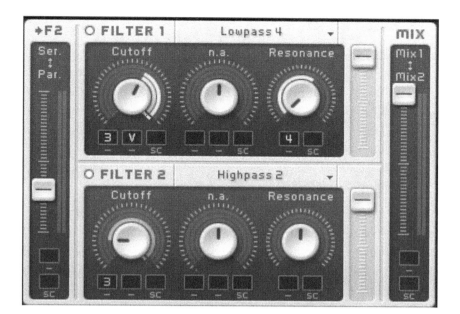

Audio Example 3.2
Explore the extensive filter options provided on instruments like Native Instruments' Massive synthesizer.

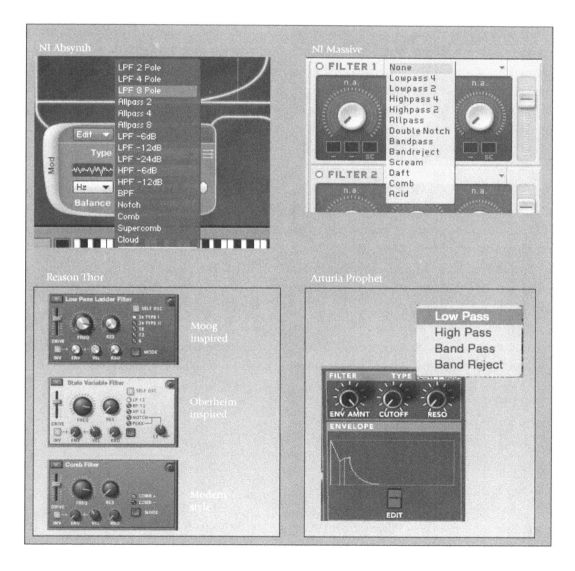

Figure 3.19
Examples of LPF/VCF sections of popular synths.

Time-Based Effects

When a skilled musician plays a sustained note on an instrument like a violin you will notice that the sound is not static; rather, it is shaped over time. Once musicians attain a certain level of technical facility on an instrument they are taught ways of making a performance *musical*. Examples of this include a regular shifting over time of the pitch center (such as vibrato), volume (tremolo), or timbre. In this section we look at tools used for creating these time-based effects on the synthesizer.

Low-Frequency Oscillator

We have already looked at the oscillator as a device that produces a user-definable waveform in the audible spectrum; a low-frequency oscillator (LFO) does the same thing but in a range that is typically below 20 Hz (and thus, not audible on its own). The frequency being produced by the LFO isn't being heard, per se, rather its shape is used to manipulate the audible sound in some way.

The most common example would be modulating the frequency of a sound in a repeating sine or triangle pattern over time, resulting in vibrato. A square-wave LFO feeding the frequency modulation of a VCO will create a British-style police siren. Modulating the amplitude of a sound with a LFO with sine or triangle waves results in a tremolo effect. And finally, at least for now, we could modulate the position of the cutoff frequency of a filter by using a LFO; doing so with a sine or triangle will result in a different kind of tremolo, one that is generally less aggressive. LFOs might also be applied to the frequency modulation of just one of two or more VCOs, resulting in a repeating phase shift, a flanger-like effect (see Table 3.3 and listen to Audio Example 3.3)

Sample-and-Hold

The waveforms in Figure 3.20 should be familiar at this point, with the possible exception of sample-and-hold (SH). With the term "sample" in the name you might be led to think this should be something from the digital era, but SH circuits go back at least to the Moog Modular synths of the 1960s. Early SH modules were purely analog, using a switch tied to a capacitor: An incoming audio signal (from a VCO or any signal sent into the module) becomes chopped into segments as the switch freezes the voltage being stored by the capacitor and then opens again at regular time intervals defined by the user.

SH can be found as an option on many digital VCOs as a timbre of it own, but more commonly on modulators like the LFO. In years past it was a popular pitch modulator for generating the sound effect of a processing computer on television and movies (see Audio Example 3.4)

Table 3.3 Common LFO Applications

Waveform	Destination	Result
Sine, Triangle	Pitch (VCO)	Vibrato
Sawtooth, Square	Pitch (VCO)	Alarm or siren effect
Sine, Triangle	Output VCA	Tremolo
Sine, Triangle	Filter cutoff (VCF/LPF)	Auto "wah"-type effect
Sample-and-Hold	Pitch (VCO)	Cartoonish computer processing

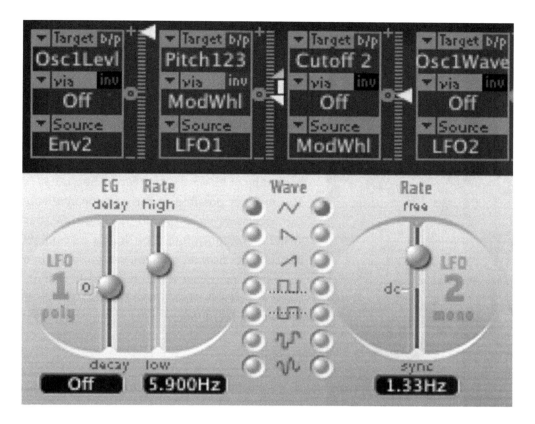

Audio Example 3.3
Explore the creative possibilities of a low-pass filter using the example of Logic's ES2 synthesizer.

Sync/Free/Retrigger

LFOs may include switches with these settings, so what do they mean? *Sync* will set the rate of the LFO relative to either the sequencer (if one is included in the instrument you are using) or to the song's tempo (if you are working in a digital audio workstation [DAW]), ensuring

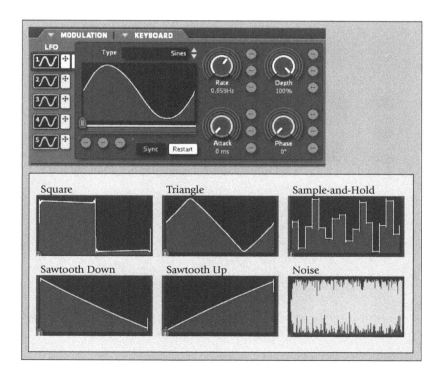

Figure 3.20
LFO modulation settings in Izotope Iris.

Audio Example 3.4

Demonstration of the Sample-and-Hold function using the ARP2600.

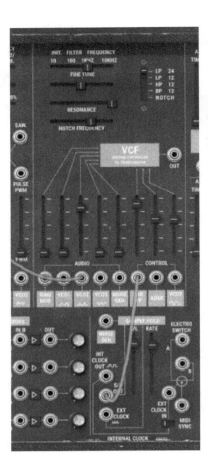

the LFO stays in a logical time division of your working tempo. When sync is selected on a software synth, the rate control might change from milliseconds to a musical value like 1/1, 1/2, 1/4, 1/8, etc., representing the time value relative to the tempo. The opposite of sync is *free*, indicating that the rate of the LFO is independent of any tempo settings.

When *retrigger* is selected, the LFO starts at the beginning of its wave cycle every time a note is struck. This was not possible on early analog synths because the oscillators were running all the time, but it's a switch commonly found on hardware and software instruments now so that you can go "old school" and leave it off or engage it and have the LFO start at the same stage each time. These controls are seen in the LFO examples of Figure 3.21.

Delay

If you think about someone playing an acoustic instrument, they normally don't start a vibrato until a note has sounded for a bit. That's exactly what this control does: It delays the onset of the LFO by the amount of time you select. On some instruments (like the Oberheim SEM V in the previous Figure 3.21) there is a "fade-in" control that doesn't delay the LFOs onset, per se, but does control the intensity of its entrance—set high, it can act like a delay.

Figure 3.21

LFO options in popular synths.

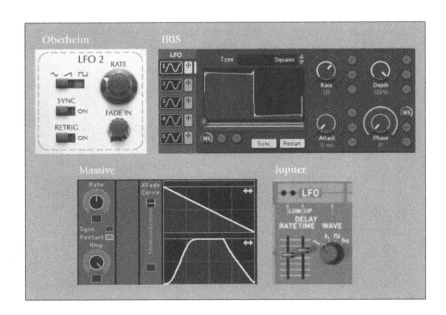

Keyboard Tracking

In some cases it may be desirable to have the rate of the LFO influenced by how high or low the notes are: faster on the high end, slower on the low end. Keyboard tracking will be a slider or knob that allows you to control how much of a variation there is.

Multiple LFOs?

It is not unusual—especially on software synths—for there to be more than one LFO available. Although they could be targeting the same variable (like the pitch of an oscillator), a great use of these extras is to send each to a different parameter. For example, a triangle wave could be controlling the pitch of an oscillator while the cutoff frequency is being modulated by a square wave. The options are endless!

Envelope Generators

How a sound's brightness and vibrato (just to name a couple) evolve over time can be manipulated manually with a controller like the modulation wheel, but they can also be defined as a characteristic of a particular patch. This is where the *envelope generator* (EG) comes into action. A typical EG provides the following controls: attack, decay, sustain and release (shortened as ADSR). In some cases it may also include hold stages in between: The H in AHDSR forces the level to remain at maximum for a specified duration of time before shifting through the decay stage to the sustain level. Figure 3.22 shows a more common ADSR envelope where attack, decay, and release controls are measured in time while the sustain portion is measured in level—its duration is determined by how long the note is held (see Table 3.4).

For percussive sounds—those that do not have sustain—all that is necessary are the attack and decay stages. Some instruments will have an envelope that includes only AD.

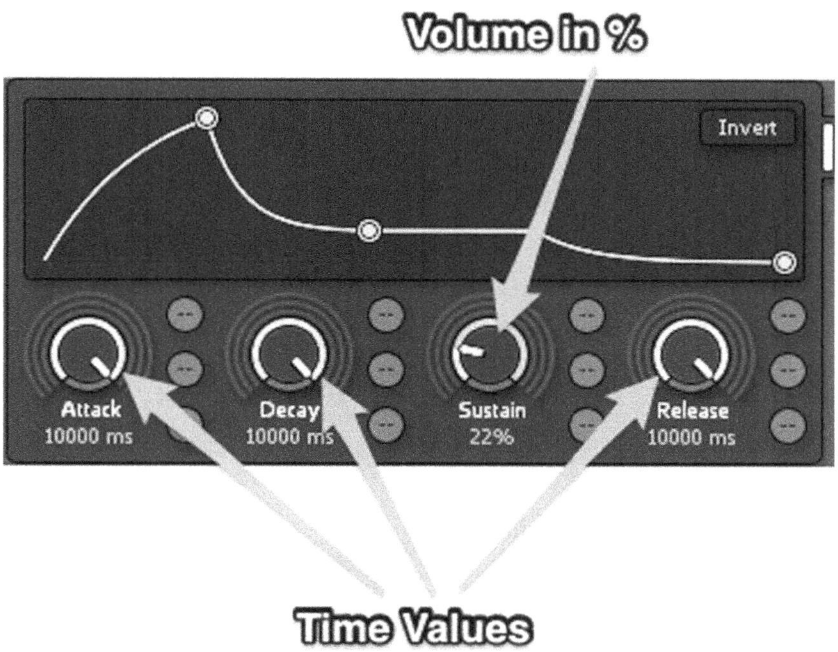

Figure 3.22

Envelope Generator stages (Izotope Iris 2).

Creating Sounds from Scratch

Table 3.4 Volume Envelope Characteristics

Attack	The amount of time it takes for a sound to go from nothing to maximum level. Percussive instruments will be short; bowed/legato instruments will be longer.
Decay	The amount of time it takes for a sound to go from the maximum level attained at the end of the attack stage to the volume *level* set in sustain
Sustain	The volume level (not a time duration) at which the sound remains while a key is held following the attack and decay stages
Release	The amount of time it takes for a sound to go from the sustain level to nothing upon release of the key

Note: On some instruments you may also see an H (hold) stage stuck in along the way. This simply allows the user to define an amount of time in which the level will be held until progressing to the subsequent stage.

If you are working with a synth that doesn't, no worries: Just turn the sustain level to 0% and the sound will end when the duration of the attack and decay stages are complete (see Figure 3.23).

The use of an EG provides control of a sound's evolution over time, a critical component that is easy for musicians using electronic instruments to neglect. Whether in emulation of a real instrument or creating something entirely artificial, a musical performance on any instrument should include a time-based variation to sustain the listener's attention and/or to evoke emotion through phrasing.

Consider, for example, a whole note (four beats in 4/4 time) with a dynamic marking of *f* (*fortissimo*) on a musical score. If we were mimicking a section of violins with a sampler, we could call up a patch that contains meticulously recorded and organized samples of a violin section. However, although the samples themselves may be very accurate, we will quickly find that striking a note on our controller and holding it down for the four beats sounds little like the real thing because it's static and not musical. The reason for this is that a good musician will vary a sound's intensity and timbre over time to build a phrase or musical arc.

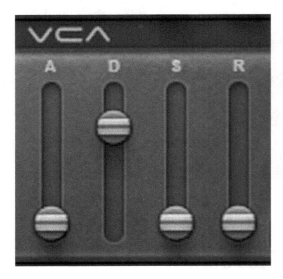

Figure 3.23
Volume envelope settings for percussion (Waves' Codex).

Even a simple shaping of the sound's volume over time will quickly give the artificial string section greater legitimacy. Exactly what shape is given will depend on the musical material that precedes or follows. For example, a *fortissimo* whole note at the end of a long and building section might call for a short and somewhat aggressive attack with a slight diminuendo (even if one isn't specified by the composer): short attack, long decay, moderate/high sustain, and a moderate release. This is very simple to accomplish with an EG. What if we want that effect for only one note in the piece, as described in our hypothetical scenario? Simple volume automation—not specifically tied to the instrument—may prove to be the most convenient solution, and software synthesizers typically allow all their parameters to be automated.

Other LFO and EG Applications

Although volume is the most obvious application here, other aspects can be controlled with an EG as well. For example, patching it to vary the cutoff frequency of a LPF to shape the brightness of a sound over time. Think about the sound of a violin being played *ff* versus it being played *pp* (very *pianissimo*). Aside from the difference in loudness, there is a timbral distinction: A violin bowed harder produces a sound that contains more high-frequency harmonics than it does when bowed with less pressure. An EG can be used to create this shape across a single note or chord.

Both the LFO and EG are incredibly useful and powerful tools when used on their own. Some synthesizers even permit one modifier to control another!—for example, varying the intensity of an LFO by modulating *it* with an EG.

When you become comfortable designing sounds from scratch, a system with lots of flexibility, like Tassman or Reaktor, is exciting because you can build the exact instrument you need. It also takes a lot of the guesswork out of understanding how all of the components are connected. That being said, Tassman in particular is a great instrument for beginners, but its sound quality and depth mean it will remain a go-to tool for sound design even as you become more advanced Let's take a closer look at the routing in Figures 3.24, 3.25, and 3.26 to illustrate just why it's so helpful for newbies, and to help

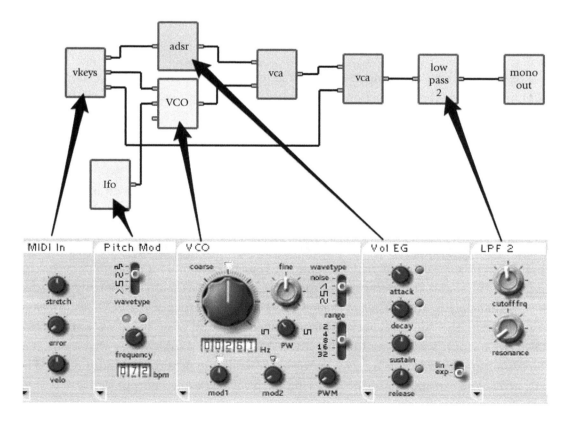

Figure 3.24
Single-oscillator synth example overview (AAS Tassman).

Figure 3.25

Detail of Tassman Builder: Stage 1.

TASSMAN INSTRUMENT: STAGE 1

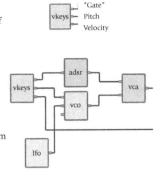

vkeys (keyboard controller information)
Gate-On/Off Signal; On sends an output of 1 volt, Off sends no output or 0 volts.

Pitch sends the note played on the controller. Octaves represent +/− 1V, so semitones are 1/12th of an octave.

Velocity sends a signal with a voltage representing how hard the key was struck.

VCO
The VCO is running at a set frequency multiplied by the Pitch voltage coming from the "vkeys" module

Its second input is frequency modulation affected by the voltage coming from the LFO

lfo
Generates a wave whose shape and frequency are defined by the user. Its signal is sent to the Frequency Modulation input of the VCO.

adsr
When a key is struck, the Gate generates an output of 1 volt.

The EG delays the increase from 0 to 1 volt based on the setting of the Attack. Once 1 volt is reached, the Decay stage begins and gradually brings the voltage to the Sustain level. When the key is released, the voltage drops again from the Sustain level back to 0 volts in the duration specified by the Release.

All of this is a patch from 0 to 1 and back to 0.

vca
The VCA multiplies the incoming signals. If no key is struck, it receives a multiplier of 0 from the Gate, and no output sounds.

Depending on the setting in adsr the multiplier will be somewhere between 0 and 1.

Figure 3.26

Tassman Builder, Stage 2.

TASSMAN INSTRUMENT: STAGE 2

vca
This VCA has a different role, it is multiplying the voltage of the incoming signal by the amount outputted by the vkeys velocity output, representing how hard the key was struck and affecting its overall volume.

low pass 2
As you might have guessed, this is a 2nd order (12 dB/octave) LPF further shaping the signal coming out of the second VCA.

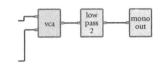

mono out
The completed monophonic timbre is outputted to the computer's sound card.

you to visualize how the components covered in this chapter integrate in a simple one-VCO instrument. The top portion of the diagram shows the "builder" side of Tassman, the bottom shows the "player."

Voltage-Controlled Amplifier

The output of a voltage-controlled amplifier (VCA) is the product of two input signals multiplied ($A \times B$ = VCA output). When working with a modular synth it may be necessary to use a VCA to produce an output from the keyboard or a sequencer. To understand this, let's consider this scenario: The VCO module is always running, always creating a voltage that, if fed to a speaker, will make a sound. If your intention is for a key on a keyboard to trigger the sound, it is necessary to turn that key into a switch. Traditionally this was done by multiplying the output of the key (1 V) with the output of the VCO. If the key is pressed down, the VCO's output will be multiplied by 1, and when the key is not pressed down, its output is zero—multiplying any value by 1, of course, keeps that value the same; multiplying any value by zero, on the other hand, equals zero, and thus no sound is outputted. You can start to see how varying the level of the key's output signal (0.0, 0.1,

Figure 3.27
Arturia ARP 2600 VCA section.

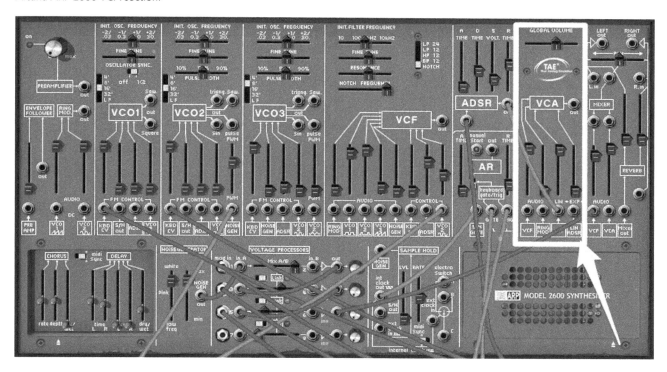

0.2 ... up to 1.0) through velocity variation (striking the key harder or softer) or when using a LFO or an EG will make the VCO's output louder and quieter through simple multiplication.

It is somewhat rare that you would come across a VCA on a digital synth, unless it is used for modeling an analog instrument (like Arturia's ARP 2600 seen in Figure 3.27). However, in visualizing the signal flow of a software synth, it can be helpful to think about the VCA's role on analog devices and how its behavior is realized in digital instruments.

Control Voltage

Control voltages (CVs) are dc signals (in contrast to audio signals that are ac) sent around an analog synthesizer as a way to share information being generated. For example, a keyboard will generate a different CV for each key and then feed it into a VCO to produce the corresponding pitch. CVs sent from an EG can be sent to a VCA and multiplied with the output of a VCO so that the delayed rise and fall of the dc voltage will dictate the shape of the sound's amplitude shape over time. The CV output of a LFO patched to a VCO is what is responsible for creating vibrato.

In terms of the actual signal, striking a key on the keyboard changes its CV output from 0 V to 1 V. The VCO's output multiplied by zero equals zero, or no sound; multiplied by 1 will output the full strength of the VCO signal. For keyboards with velocity control, the range of velocity can be output as somewhere between 0 and 1 V; thus the multiplier will define the intensity of the VCO's output. This is referred to as the

Creating Sounds from Scratch

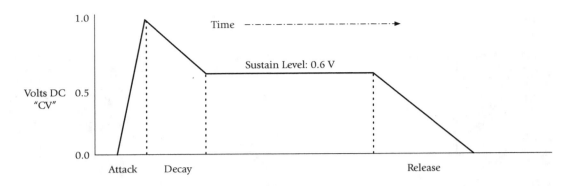

Figure 3.28 Diagram of control voltage values through an EG—Attack, Decay, and Release represent time values; Sustain is a level value with its duration determined by the performer.

gate (you might also see the term *trigger*): "0" means the gate is closed, "1" means the gate is wide open.

In the case of the EG, its output of 0 V to 1 V is ramped up for the duration of time defined in its attack stage, then it drops from 1 V in the decay stage to something lower (say 0.5 V) at the specified sustain level, and when the key is released, the release time defines the duration back to 0 V, at which point the VCA is again multiplying the audio signal by 0 V and there is no output. These values can vary between manufacturers, but 0–1 makes for easy math and a convenient way to understand how the CV signals do their thing (see Figure 3.28).

Prior to the advent of the MIDI, it was not uncommon for electronic music studios to have multiple synths, each with inputs and outputs labeled for CV signals. The gate signal of one keyboard could be patched to VCAs on other instruments so that they all play in unison. A LFO on one instrument could be routed to another and control the same corresponding parameter or something entirely different—like frequency of a VCO on one instrument and cutoff frequency of a LPF/VCF on another. The same is true for EGs. You might say you are limited only by your own creativity in these scenarios!

When MIDI instruments came on the scene, CV inputs/outputs (I/Os) became unnecessary and went away because that information, and much more, could be transmitted easily in the form of digital data. However, with the rekindled interest in analog synths, CV is showing up on new instruments, and MIDI-to-CV converters—which have been around since the early days of MIDI—are still available for bridging the gap between these otherwise incompatible protocols.

Modifiers: Special Effects

Distortion

By definition, distortion is altering the shape of the original waveform. Figure 3.29 shows a pure sine wave on the top followed by the waveform resulting from various means of distortion.

A good place to start is with the first example, which is simply overdrive, meaning the signal was increased in gain beyond the limits of the signal path. Notice how the top and bottom of the waves flatten out? The sine wave becomes distorted to the point that it begins to look more like a square wave. What does a square wave have that a sine wave does not? If you said "harmonics," you got it! Increasing the amplitude of a sine wave beyond the

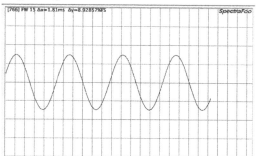

Figure 3.29
Examples of distortion applied to a sine wave.

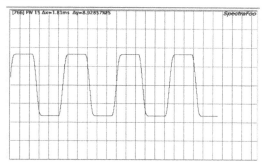
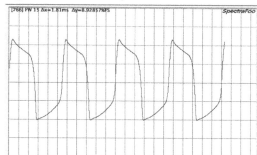
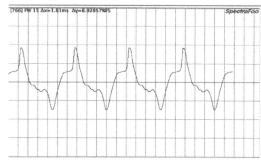
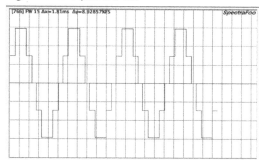

limits of your signal path will cause *harmonic distortion*, altering the original waveform by adding odd-numbered harmonics—the very ones that make up the harmonic structure of a square wave.

The other examples include the same sine wave processed with the "Nasty" setting in Logic's Distortion II plug-in, a modeled Vox AC30 guitar amplifier, and Logic's Bitcrusher plug-in that reduces the bit rate of the inputted signal to create digital distortion.

Delay-Based Effects

This category incorporates a lot more effects than you might think. Echo is perhaps the most obvious, but it is a mistake to think of delay effects as solely those that produce a discernible repetition of the dry sound. When the inputted sound is repeated within a time frame of less than around 50 ms the impact is more timbral in nature—meaning our hearing system may not be able to detect two sounds and fuse them together. Let's go through some of the more common variations of delay-based effects used in synthesis.

Creating Sounds from Scratch

Echo

With a delay time set to ≥50 ms a distinct repetition of the original sound is audible. The earliest version of this effect was accomplished with reel-to-reel tape machines that had separate heads for erase, record, and playback. The dry signal would be sent into the machine and printed to tape through the record head, and then shortly thereafter (depending on the tape speed) reproduced back into the mix through the playback head. The timbre of the echo is colored by the tape-recording process, but even in digital it is common to filter the repeated sound to distinguish it further from the original signal. With analog setups it was common to feed the delayed audio into the console and use a channel EQ or filter. With echo or delay plug-ins it is common for there to be at least a LPF, as the attenuation of high frequencies in the repeated sound distinguishes it from the original and also mimics the loss of high frequencies inherent in a repeated acoustic sound lost to absorption in the reflecting surface and resistance in the air.

Another common echo trick was to "bleed" a little bit of the delayed signal returning to the console *back* into the tape machine to generate a controlled feedback loop. Feedback now is a control that is available on nearly all digital delay pedals and plug-ins.

Delay or echo is commonplace with vocals, but is helpful for adding depth or space around any sound. It can be configured with repeated sounds falling on rhythmic values linked to the song's tempo. Figure 3.30 shows the standard delay plug-in that comes with Pro Tools, which includes these parameters and more.

Reverberation

Whether the goal is to simulate a real acoustic environment or invoke a decidedly unnatural decay on your timbre, "reverb" is the place to go. There are three major categories of reverb: acoustic, mechanical, and digital.

Acoustic Reverb: Halls, Studios, Cathedrals/Churches, Chambers Acoustic reverb is reverb generated from an actual space; in synthesis this would mean sending your sound through a loudspeaker into a concert hall or studio and capturing the resulting sound with

Figure 3.30 Pro Tools delay plug-in: Mod Delay III.

microphones (one for mono, two for stereo, five or more for surround). Before mechanical reverberators came along, large studios had dedicated reverb "chambers" (sometimes called "echo" chambers, although they were typically too small to produce an actual echo) to provide a natural sense of ambience in the mix. Chambers come in all shapes and sizes, often dictated by whatever space in the facility is not otherwise being used! There are home studios with converted wine cellars (questionable priorities, perhaps) or garages that work well as chambers. Ideally there should be no parallel surfaces so that standing waves can be avoided, but that is not always practical. Diffusors and baffles around the room can help customize the decay time and color and minimize the effect of standing waves. A freely movable speaker and microphone(s) add further flexibility and room for creativity. Figure 3.31 provides a hypothetical example with visible plumbing pipes (in gray) acting as diffusors, much like the original configuration of the famous chambers of Abbey Road Studios in London.

Although true chambers tend to be the domain of facilities rich in space, electronic sounds benefit greatly from passing through a decent speaker, into a quiet room of any kind, and picked up with a pair of directional microphones facing away from the speakers (to minimize the amount of direct sound captured). This adds a space and depth around an artificially generated timbre that is not as convincing with artificial reverbs. Even if your studio is a bedroom, try sending one of your tracks out your monitors and back to a stereo audio track with a couple microphones (of course, be sure not to route your microphones back to the speakers *yet* or the intense feedback may send you searching online for new monitors and a nice pair of hearing aids).

Mechanical: Plates and Springs With the absence of a suitable acoustic space prior to the ubiquity of digital reverbs, mechanical reverbs offered the next best thing. Far from producing a realistic reverb, mechanical reverbs generated a decay that could be controlled in duration through mechanical damping and in timbre with EQ and filters.

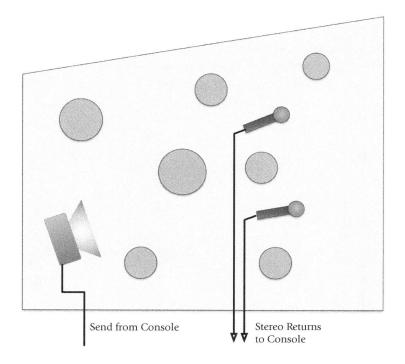

Figure 3.31
Chamber reverb diagram: speaker facing into reverberant room with column diffusors, and two microphones returning the sound to the mix.

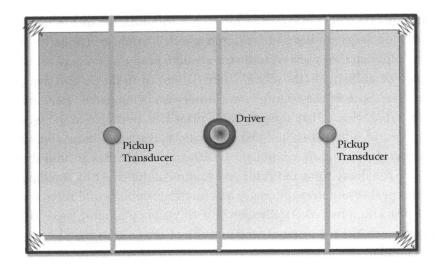

Figure 3.32
Plate reverb mechanism.

A plate reverb is constructed of a large piece of sheet metal (the larger it is, the more low-frequency resonance) that has been connected to it what is effectively a loudspeaker driver vibrating with a signal fed from the mixer, resulting in complex vibrations within the plate. Two (for stereo, one for mono) transducer pickups are placed elsewhere on the plate to transfer the vibrations back into the mix. All of this is suspended with springs inside a wooden or metal enclosure and, ideally, stored in a quiet area (so as to avoid sound in the room causing unintended vibrations in the plate).

There are some very convincing digital emulations of plate, the best of which capture the changes in the plate's behavior when fed a low-level signal versus one that is very high (see Figure 3.32). As you might imagine, a "loud" signal coming in will cause far more complexity in the vibration patterns, perhaps even distorting, that may be just what you are looking for, regardless of whether it is to be blended at a high or low level in the mix. Another layer or creativity often overlooked!

Spring reverbs have been installed in guitar amplifiers for many years. If you have ever dropped or bumped a guitar amp with it turned on, you may have heard a horrendous sound like someone whacking a spring that has been amplified—well, that's what it was! A simplified version of the mechanism is diagrammed in Figure 3.33. One end of the spring is connected to a driver that vibrates with the inputted signal; the other end is a pickup that brings the decaying sound in the spring back into your mix. Although more

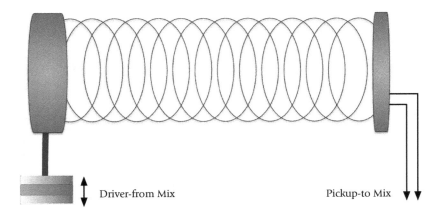

Figure 3.33
Spring reverb mechanism.

commonly found on amplifiers, there were spring reverbs in rack-mount form and as large stand-alone boxes that were used in recording studios. Some of these still exist and, of course, there are several good plug-in options available.

As we said, although the mechanical devices do not emulate a space, engineers and musicians found their favorite applications for the quality of decay they create, sounds that are still desirable today. Plates are especially good for drum sounds; sending a good-quality snare sample through a plate reverb really brings the sound to life in a familiar way. Springs are most associated with guitar amplifiers and are thus a popular option when one is using electric guitar samples or recording a guitar directly to a DAW with amp simulation.

Digital Reverbs: Algorithmic and Convolution And finally the category that you are likely already acquainted with, digital reverb. Because the acoustics of a real room comprises a series of reflections whose timbres are colored by the physical materials they are bouncing off of and interactions with each other, digital delays can be programmed and manipulated to simulate convincingly real spaces or those that are intentionally artificial (see Figure 3.34) . In fact, this category can even be seen as a catchall for reverb in general in that the categories of nearly all devices include settings for hall, room, studio, chamber, plate, and spring. See Table 3.5 for a description of the most common parameters on digital reverbs.

One final category of reverb, before we move on, begins to overlap actual synthesis techniques, and that is convolution. The goal of convolution is to capture a reverb—whether it be acoustic, mechanical, or digital—and merge it with your dry audio signal, effectively placing your dry audio signal into that effect. It can be a great way to put your sound in a cathedral or the live room of a famous recording studio, or through a classic plate or spring reverb.

Figure 3.34
Lexicon
LXP Native
Digital Reverb.

Table 3.5 Common Parameters on Digital Reverb (Lexicon LXP Native as example)

Waveform	Destination
Reverb Time (RT60)	Duration of the reverb delay; sometimes described as the time it takes for the decay to attenuate by 60 dB
Rolloff	Position of high frequency attenuation (LPF)
Diffusion	Spacing of reverb reflections (generally high for percussive sounds, low for legato/sustained sounds to minimize masking)
RT HICUT	Reverb's high-frequency cutoff, helps describe the density of the surfaces in the artificial space being crafted
MIX	Reverb balance: 0% has no reverb, 100% is only reverb

Convolution reverb can be as simple as calling up a plug-in like McDSP's Revolver (Figure 3.35, left) or Audio Ease's Altiverb (Figure 3.35, right), and calling up a preset of a room or device you want to use. Because you are "convolving" your sound with an "impulse response" prerecorded as a preset, the parameters are limited compared with those of a reverb based solely on algorithms. Typically you have control over the mix percentage of wet to dry signal, control of the decay time (less than or equal to the sampled duration), predelay, and EQ and/or filters. So customization isn't exactly a strength, but realism most certainly is. Early convolution reverbs sounded great but had very limited options for customization, although that has changed with newer plug-in versions.

Applications like Logic come with a utility (See Figure 3.36) for creating your own impulses, and it's a lot simpler than you might think. The utility will generate a sound (sine-wave sweep from 20 to 20,000 Hz, is common) that is fed to a loudspeaker in a room or inputted into a reverb device. Either with microphones (two for stereo, five or more for surround) placed appropriately in a space or with the output of a reverb device fed back into the utility, the software "deconvolves" the sine-wave sweep from the captured recording and creates an audio file that is loaded into your convolution reverb plug-in. There are *lots* of impulse responses (IRs) online that you can download and try out in your plug-in of choice.

An often overlooked creative use of convolution reverb is putting aside your nicely deconvolved reverb samples and instead using some other audio sample like a voice or instrument recording, thus merging that sound with your audio signal. It's simple to do and can yield interesting results. Also consider capturing reverb-*like* sounds, such as feeding the sine-wave sweep through a loudspeaker into the hole on an acoustic guitar, and then using a microphone or two to capture the resonance of the guitar's body and

Figure 3.35 McDSP Revolver and Audio Ease Altiverb 7 convolution reverbs.

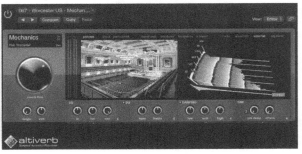

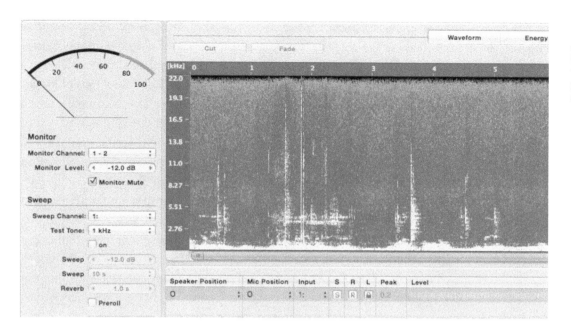

Figure 3.36
Apple's Impulse Response Utility for creating your own reverb samples.

vibration of the strings. Do the same into the open lid of a grand piano with someone holding the sustain pedal down (treat them to a pair of earplugs for their efforts).

Phaser

The phaser effect reintroduces an audio signal back once or several times on top of itself at a user-defined delay. The process is a creative application of comb filtering—something that was once seen as only a problem to avoid—in which peaks and dips are created in the frequency spectrum as the delay causes some frequencies to meet constructively, resulting in an increase in intensity, whereas others meet destructively, causing an attenuation. Figure 3.37 shows Blue Cat's phaser plug-in that provides easy-to-understand controls. Notice the feedback control that spills some of the output back into the phaser to intensify the effect. An LFO is commonly used to shift affect the delay and depth in a repeating pattern—notice here the shape of either a sine wave (at the bottom) or a triangle wave (at the top) of the switch on the right-hand side. The allpass filters we already covered are a key component in a phaser design.

Chorus and Flanger

Next to reverb, chorus and flanger are the most commonly used delay-based effects. Guitarists like to have one or both on their pedalboards; keyboardists use chorus on a lot of presets to make the sound deeper and more colorful.

Figure 3.37
Blue Cat Phase.

Creating Sounds from Scratch

The term "chorus" comes from the idea of trying to make one voice sound like an entire chorus. You are not likely to fool anyone with that trick, but it a useful effect nonetheless. The sound is achieved by layering several delays within a range of 20–50 ms, resulting in comb filtering but of a different quality than you get from the phaser. The different stages are slightly detuned, adding to the complexity and mimicking the lack of tuning precision that is not just impossible but desirable in a section of violins, voices in a choir, or the keys of a piano that have two to three strings that are not 100% in tune with each other. Chorus examples that might help place the sound in your mind's ear: the clean guitars in Nirvana's "Come as Your Are" or the electric piano on The Eagles' "I Can't Tell You Why."

With a name deriving from its origin, a flanger effect was first created with two tape machines: By placing a finger on the outside of the reel (called the "flange") on one machine, its speed would slow down relative to the one left untouched, thus affecting a gradual slowdown that took the pitch down and created a shifting comb filter when both signals were mixed together. When the finger was released, the tape sped back up, and the two outputs once again matched. There are countless examples of this effect being used over the years but it was heard frequently in the 1960s on recordings of the Beatles and Jimi Hendrix, just to name a couple.

Sometimes it's difficult to clearly identify chorus vs. flanger because they have similar characteristics, and that is why they are grouped together here (see Figure 3.38). In general, chorus is used more gently to enhance the stereophony of a sound, and flanging is used more aggressively—but there are clearly exceptions to this statement.

Ring Modulation

Ring modulators have been around since the early days of synthesis and work on the principle of amplitude modulation. In amplitude modulation, two signals are combined in such a way that the output consists of the sum and difference of two waves. In other

Figure 3.38 Blue Cat Flanger and Chorus.

Figure 3.39
Ring modulation effects: Logic and Arturia Moog Modular.

words, if we modulated a 200-Hz sine wave with a 300-Hz sine wave, we would end up with one at 100 Hz and another at 500 Hz, in addition to the original 200 and 300—not terribly interesting with sine waves but quite so when using complex waves made of many sine waves at different frequencies across the spectrum.

Commonly referred to as "ring mod" in the jargon of the studio, the signal you input will be combined with either a wave generated within the processor or with a second signal you provide through a side-chain input. Figure 3.39 shows Logic's Ringshifter and the Ring Modulator module found on the big Moog (Arturia's plug-in version, in this example). The results can be subtle with a little phase distortion or extreme and metallic sounding. Fans of Doctor Who will recognize the sound as the effect on the voice coming from a Dalek. Suffice to say, the results of the effect move quickly across a spectrum of mellow … to jarring … to downright painful!

Tuning—Fine vs. Coarse

This one seems as though it should be obvious, right? Maybe, maybe not. Oscillators will have pitch control for the basic function of tuning with another instrument or recording that's not quite in tune, but it has a creative application as well. Coarse tuning normally shifts by a semitone up or down, which could be used for playing in one key but hearing it in another, or detuning one oscillator relative to another to play an interval with each note (unless the interval is an octave, watch out for notes that may fall outside the scale being played; harmonizers often have the facility for defining a scale so that accidentals are added as necessary). Fine tuning, on the other hand, is a slight detuning of one oscillator against the other in *cents* (100 divisions of a semitone), enabling a chorus-like effect when closely detuned against another. Tying an LFO to a fine-tuning control creates a pretty convincing chorus oscillation. Figure 3.40 is a detail of the Moog Voyager's oscillators 2 and 3, which have both fine tuning (top) and coarse octave tuning (middle).

Creating Sounds from Scratch

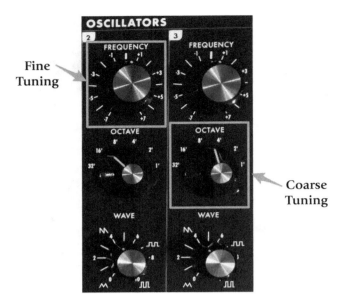

Figure 3.40
Tuning controls from Minimoog Voyager (Photo: Metcalfe).

Unison/Voice Stacking

Unison mode (sometimes called voice stacking) has its origin in polyphonic analog synths as a means to fatten up a lead or bass tone. Two or more (depending on the instrument's capabilities) voices sound simultaneously, but, because of the inherent pitch instability of analog oscillators, the resulting phase distortion adds a complementary color. Digital synths emulating the imperfections of analog may have a unison option with a pop-up menu for choosing the number of voices to be layered (see Figure 3.41).

Portamento/Glissando

In music, *glissando* is a performance instruction from the composer or orchestrator asking the performer to go from one note to another, hitting every note in between. On a fretted instrument, like a guitar, this would mean steps of semitones; on a piano this would mean running a finger up the white or black keys. On a nonfretted instrument like a violin, the transition needs no such restriction. That is also the case

Figure 3.41
"Unison" voice stacking on Arturia Jupiter 8V.

with a synthesizer bcause the oscillator (or other sound generator) can smoothly shift from one note to another, hitting every frequency along the way. This is called *portamento* (you might also find it labeled as "glide" on your instrument). As seen in Figure 3.42 there will be a control for how much time it takes to get from one note to the other.

Figure 3.42
Portamento/glide settings on Logic's ES2 and Arturia's Jupiter 8V.

It's interesting to note that the behavior of *portamento* will be different depending on whether you are in monophonic or polyphonic mode: in mono, the result is what was already described; in stereo, the effect can be very dramatic with each note of a chord chasing new notes when there is a chord change. This is a worth a little demo, as Audio Example 3.5 on the Companion Website provides.

Audio Example 3.5
Applying portamento (also known as "glide") on a Minimoog patch.

Modulation Matrix

The modulation matrix can come in different forms, but one of the most visually intuitive and incredibly powerful one is found on Logic's ES2 (see Figure 3.43). Rather than a simple switch that allows the user to set a LFO, say, as a pitch control (for vibrato) or amplitude control (for tremolo), a matrix allows the user to choose a source (like a LFO) and a destination (like pitch) with the option of an intermediary step. For instance, the first column in this example shows the modulation wheel ("ModWhl") acting as the gate through which the LFO must pass before it can affect pitch. On the right-hand side of the column is a two-part slider defining the upper and lower intensities of the source: In this case the lower is slightly above the middle position, which is zero, and the upper is only partway to the top of the scale. Therefore, when the modulation wheel is all the way down, there will still be vibrato with a narrow pitch range, and when the wheel is rolled all the way up, the pitch range will be much greater. The second column is a simpler routing that gives full control (notice the single slider all the way at the top of the scale) of how hard the controller is played (velocity) and how bright the resulting sound is (cutoff frequency of both VCOs 1 and 2).

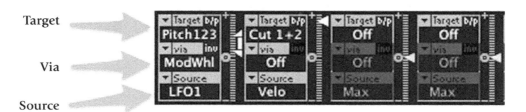

Figure 3.43
Modulation matrix from Logic ES2.

Figure 3.44 C major arpeggio in common arpeggiator patterns.

Up Pattern   Down Pattern   Up/Down Pattern

Arpeggiator

In music an arpeggio is defined as a chord that is broken into its individual pitches. An arpeggiator is a synth tool for automatically breaking a chord into its individual parts and following a user-defined or preset pattern. Figure 3.44 illustrates common patterns with a Cmaj7 chord. The player enables the arpeggiator mode, selects a pattern (shown are the common Up, Down, and Up/Down patterns, and hold the chord C–E–G–B. The arpeggiator spells out the chord in the selected pattern at a user-defined tempo. It may seems like cheating—and could be at times—but it's also a great way to get interesting, dynamic textures that would otherwise be unplayable, especially in a tight quantized pattern at a fast tempo.

Figure 3.45 shows the arpeggiator section of the Jupiter 8V synth highlighted. The top-left section shows a slider for "Rate" and a switch for an internal clock or locking to an external clock. On the bottom right, in addition to Up, Down, and Up/Down ("UD") there is an option for random ("RND") that, as you may have guessed, randomly selects the next note in the pattern. The instrument also enables storing four different patterns (1, 2, 3, and 4) for recall when needed.

Pattern Sequencer

A step up from the arpeggiator, so to speak, is the pattern sequencer (see Figure 3.46). Not to be confused with a song sequencer (like you would use recording in Logic, Cubase, or DP, to name a few), the pattern sequencer is used to play a predefined pattern of notes each time a note is played on the controller. This was at the heart of the keyboard-less Buchla instruments but may also be found on instruments with a keyboard.

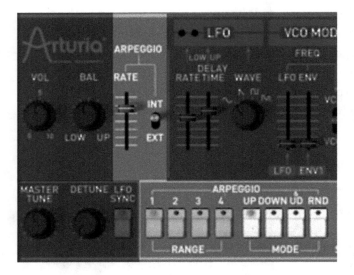

Figure 3.45 Arturia's Jupiter 8V arpeggiator section.

Oscillator Sync

When enabled, the oscillator sync hard-synchronizes the start of each wave cycle on one oscillator with another. Figure 3.47

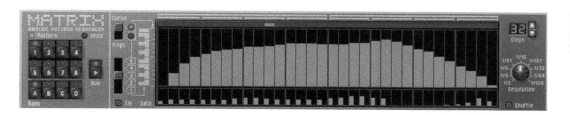

Figure 3.46
Reason Pattern Sequencer.

shows the two VCOs on the Oberheim SEM V. The arrow in the middle is pointing to the sync switch. VCO 2 is acting as the "master" and VCO 1 is the "slave." Independent of frequency, when the wave cycle in VCO 2 starts over it forces VCO 1 (regardless of where it is in its own cycle) to start over again.

Pitch Bend/Modulation Wheel

Perhaps the two most common controls on a keyboard controller are the pitch bend and the modulation wheel (see Figure 3.48) located to the left of the keys. As we explore later in this chapter, both controls are simply outputting MIDI data and can be routed anywhere. However, pitch bend is usually dedicated to pushing the pitch up or down, giving the keyboardist a means of pitch manipulation that is not possible on analog keyboard instruments. Software and hardware instruments will have a setting for how wide a range in semitones the pitch bend covers.

With the exception of many Roland keyboards, the modulation wheel ("Mod Wheel") moves freely with no resistance and stays in place when released. (Roland likes to combine the two controls onto a single joystick so that the x axis is pitch bend and the y axis is a spring-loaded mod wheel.) The importance of the mod wheel is clear in the MIDI Implementation Chart: Controller #1! It can be assigned to any destination in an instrument but is most commonly tied to regulating the amount of vibrato. It is also useful for positioning the cutoff frequency of the LPF. It is a great way to manage the volume level of string, wind, and brass instrument orchestral patches: your left hand can shape the phrase while your right hand is playing the music line (a feature first used in the Garritan Personal Orchestra). Figure 3.48 shows the pitch bend and mod wheel on a Yamaha Motif 8 keyboard.

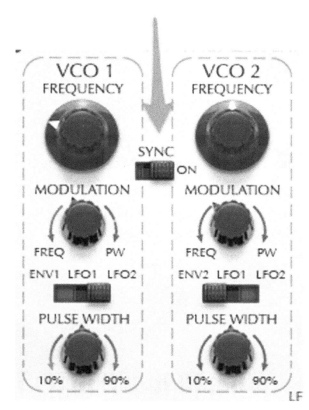

Figure 3.47
Oscillator sync switch on Oberheim SEM.

Creating Sounds from Scratch

Figure 3.48
Common Pitch Bend and Modulation wheel configurations.

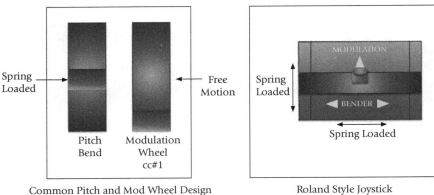

Instruments

Up to this point we learned about each single component of a synthesizer, taking basically a modular approach to explaining what really makes a synthesizer. Although it is crucial to be able to recognize its module's function, with modern synthesizers you rarely will interact with single components, but you will instead be accessing each module's function in a more integrated way. The integration of all the modules and components we discussed so far takes shape into is what we call a *digital hardware synthesizer*, sometime referred also as a *MIDI synthesizer*. The hardware synthesizer can have different characteristics in terms of type of synthesis, shape, features, and functions. Parameters such as polyphony (number of voices that can be played simultaneously), presence or lack of a physical keyboard, number of keys, presence of a MIDI interface, additional physical controllers (pads, faders, knobs), and number of outputs determine the functionality and specifications of the hardware synthesizer. It is very important to choose the right devices and instruments to use in your studio. Remember that they are the "virtual musicians" that will be featured in your music productions, so it is essential to have the right type of equipment, the right variety of instruments, and a very flexible and versatile palette of sonorities to choose from in order to be an all-round electronic producer.

The Digital Hardware-Based Synthesizer

Digital hardware synthesizers (Figure 3.49) integrate three main components: a MIDI interface (we describe in detail the MIDI protocol a bit later in this chapter), an internal digitally controlled sound generator, and a physical keyboard. If they come equipped with a built-in sequencer, then the term MIDI *workstation* is more appropriate, because they can be used as stand-alone MIDI production studios. The digital hardware synthesizer is probably the device you are most familiar with. It is also the most complete, because it can be used as a MIDI controller,

Figure 3.49
The Digital MIDI synthesizer.

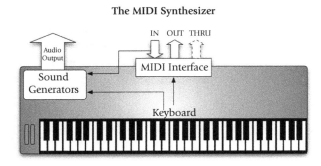

therefore allowing you to control an external MIDI device through its keyboard, but can also produce sounds through its internal digital sound generator.

Notice how the three elements are connected to one another. The keyboard sends signals (according to which note you pressed) to both the MIDI OUT and the internal sound generator, which also receives MIDI messages from the MIDI IN port. A very common, and more modern, version of the hardware synthesizer can feature a built-in MIDI host that connects directly to your computer via a USB connection (Figure 3.50). This configuration has several advantages such as limiting the number of cables running in your studio, speedier transmission of the MIDI data between the synthesizer and your computer, and the possibility of directly editing the parameters of the device from a connected computer through the use of dedicated software.

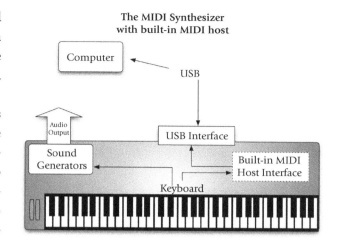

Figure 3.50
The Digital MIDI synthesizer with built-in MIDI host.

Although in the past this category of MIDI instruments constituted the core of a production studio, these days hardware synthesizers play a bigger role in live performance settings than in studio work. They have been replaced almost entirely with the combination of MIDI keyboard controllers and software synthesizers. The MIDI synthesizer is still an extremely useful and practical device for live situations, where portability, features, and reliability are key factors.

The Keyboard Controller

A modification of the MIDI synthesizer is the keyboard MIDI controller (Figure 3.51). This device features only a MIDI interface (usually only a MIDI OUT) and a keyboard. There's no internal sound module. In fact it is called "controller" because its only use is to control other MIDI devices attached to its MIDI OUT port. As it is the case for the MIDI synthesizer very often you will use a MIDI controller that has a built-in USB host and that allows you to connect the controller directly to your computer running the DAW or software synthesizers of your choice.

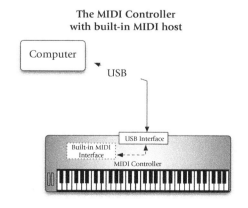

Figure 3.51
The MIDI keyboard controller.

The Sound Module

To be able to expand your sound palette you have a few options. In what now seems a distant past, you could get a new (or several new) MIDI

Creating Sounds from Scratch

Figure 3.52
The Sound Module.

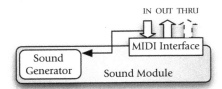

synthesizers with the latest sound engines and new patches available. This option had the problem that you would end up with several keyboards in your studio that most likely you wouldn't utilize because you just needed their new synthesize engines. To overcome this issue manufacturers started creating what we call *sound modules*. Because in most project studio situations one MIDI controller is sufficient, a sound module (Figure 3.52) features only a MIDI interface and a sound generator but not a keyboard. The advantage of this type of device is that it delivers the same power as that of a MIDI synthesizer in terms of sounds but in a more compact design and at a lower price.

Again, as was the case for the hardware synthesizer, nowadays we mostly use software synthesizers to expand our sound palette. This is a much more versatile and, in same cases, inexpensive solution.

Software Synthesizers

Software synthesizers have practically become the principal source of sounds in the modern project studio. In this setting most of the hardware gear has been replaced with software-based synthesizers. The main idea behind this approach resides in the fact that modern digital hardware synthesizers are nothing more than a basic "computer," with a dedicated central processing unit (CPU) inside and a software specifically written for that CPU that runs on it (Figure 3.53). Software (and hardware) music companies started to take advantage of the powerful and highly versatile CPUs of the personal computers to generate sounds by simply writing software that would run on these CPUs (Figure 3.54). In general this approach is much cheaper. Even including the initial cost of a desktop computer, in the long run it is cheaper to buy software synthesizers. Software synthesizers are much more flexible, versatile, and upgradable than their hardware counterparts.

Figure 3.53
The hardware synthesizer.

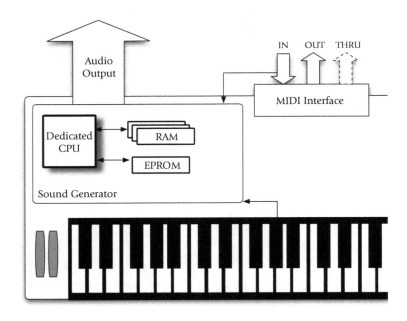

Another big advantage is that they are much easier to program because you can do all the editing by using the big screen of your computer. Even in live settings these days, laptops and software synthesizers are rapidly replacing MIDI hardware synthesizers.

One of the major drawbacks of the software synthesizer is is that you need a powerful computer and a large amount of RAM to keep up with multiple software synthesizers running inside your DAW. In addition, with the CPU requirements of software synthesizers increasing exponentially every year, you might find yourself in need of a new computer sooner than expected in order to keep up with the latest version of your favorite soft synth.

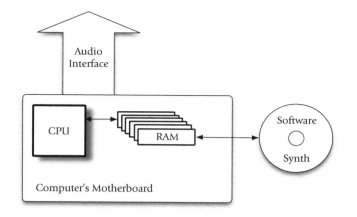

Figure 3.54
The software synthesizer.

Instead of buying a hardware box from your local music store, in the case of a soft synth you buy only the software that will be installed on your computer through either a boxed CD/DVD or through a simple download (the latter is becoming more and more popular among distributors). Depending on the type of product you bought (types of synthesis, quality of sounds etc.) the installation size taken by a soft synth on your hard disk (HD) can vary from a few hundred megabytes (for such synthesizers than are not sample based) to several tens of gigabytes (as in the case of professional orchestral and acoustic instrument libraries). A software synthesizer usually is made of two separate components: the actual software application (plug-in and/or stand-alone components) and the library (program, samples, or waveforms). We strongly advise to install these files on one or more dedicated HDs. If you plan to share your libraries among several computers (like a desktop and a laptop) you should install the libraries on an external HD with a fast connection such as USB 3, eSATA, USB-C, or Thunderbolt. Once you have installed your software synthesizer, you usually have the options to use it in either "stand-alone" mode or in "plug-in" mode. The main difference between the two modes is that the former doesn't require any host application or DAW to run and it runs independently from any other program in your computer, whereas the latter requires your DAW to be launched and running in order for the soft synth to be launched as a plug-in (a plug-in is a way of extending the features of a host program, in your case your DAW). We recommend using the stand-alone version for either patch editing/creation, or for live performances, whereas the plug-in version should be used for sequencing, writing, and producing (Figure 3.55).

Because of the relatively low cost of personal computers, it is very common these days to use more than one computer in your studio. Although one computer will run the DAW and some software synthesizers (these will serve as the "master" computer), the other(s) serves as additional sound sources (satellites), all running a series of soft synths of their own. Basically the satellite computer(s) replaces the old-fashioned and less powerful sound modules with an incredibly flexible and sophisticated sound source (Figure 3.56).

Creating Sounds from Scratch

Figure 3.55
An example of a software synthesizer (Waves Codex) used as a plug-in in Logic Pro X.

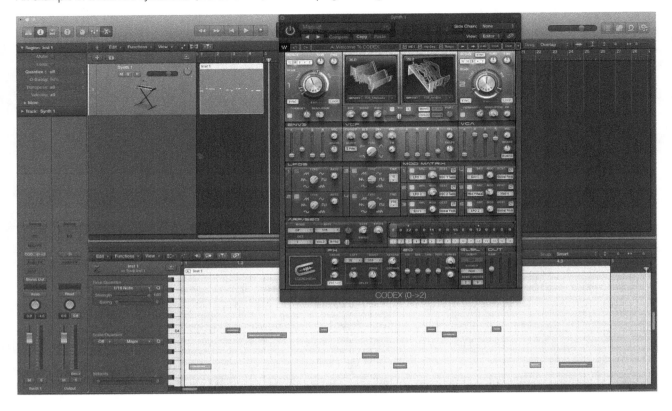

Hybrid Synthesizers

Hybrid synthesizers are not among the most common devices found in a studio but have a special niche for complex live settings that require intricate and advanced patches. They are basically portable computers (most of them running a highly customized version of Linux OS) encased in a portable rack-mountable case of two or more units. They offer the same power and flexibility of a computer running soft synths through plug-ins. They are sturdier and more portable than a full desktop computer. One of the most popular devices in this category is the Receptor line by Muse Research (Figure 3.57).

Figure 3.56
Production studio based on networked computers.

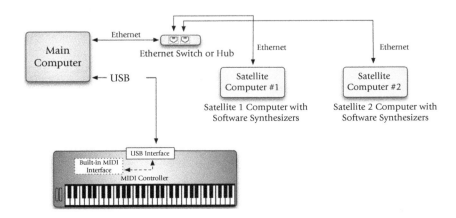

Figure 3.57
The Receptor 2+ by Muse Research.

The specifications of the Receptor (Dual-Core Intel 3.3-GHz processor, 2-TB hard drive, 8 GB of Ram, etc.) suggest that it is a full desktop computer but the exterior look makes it seem like a regular sound module.

The MIDI Standard

The Musical Instrument Digital Interface (MIDI) protocol is a communication standard developed in the early 1980s that allows different devices, made by different manufacturers, to exchange data and communicate with each other. The protocol was originally created by engineers from Sequential Circuits in 1981 and further developed by a joint collaboration of other popular manufacturers such as Roland, Yamaha, Korg, Oberheim, and Kawai. The official ratification was signed in 1983.

The MIDI protocol is based on a serial interface that runs at the speed of 31,250 bits/s. Being a serial protocol means that any type of data transferred on a MIDI cable is sent one after the other. Even though this may seem a slower way of communicating, compared, for example, with that of a parallel system in which several lanes of data can be sent simultaneously, it is in fact more reliable. MIDI data are based on a binary system in which each piece of information is sent in the form of 1s and 0s (bits). To make the data transferring more efficient and organized, bits are arranged in a group of eight, forming a "byte." Two or more bytes form a "word" (Figure 3.58).

MIDI data have the advantage of being very small and concise. A typical MIDI message, for example, is the Note On. This message is sent from a MIDI controller every time we press a key. These particular data indicate the note pressed (Note Number) and how hard that key was pressed (Velocity). Keep in mind that only the description of the note played is sent over MIDI, and not the sound itself. This is the main reason why MIDI is so "light" in terms of data transferred. The receiving device (a sound module, for example) will receive the data and pass them to the internal sound generator that only then will produce the sound based on the note number and the velocity received. MIDI is a bit like writing notes on paper (notes and dynamics or note number and velocity); the musician that will play the notes written in the parts is the sound generator.

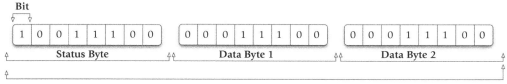

Figure 3.58
The binary system: bits, bytes, and word.

Table 3.6 MIDI Messages Categories

Channel Messages	System Messages
Channel voice: note on, note off, monophonic aftertouch, polyphonic aftertouch, control changes, pitch bend, program change	System real time: timing clock, start, stop, continue, active sensing, system reset
Channel mode: all notes off, local control (on/off), poly on/mono on, Omni on, Omni off, all sound off, reset all controllers	System common: MTC, song position pointer, song select, tune request, end of system exclusive
	System Exclusive

When the MIDI standard was ratified, the manufacturers agreed on a standard codified data structure that all MIDI devices would understand. MIDI data were divided into two main categories: Channels and Systems (Table 3.6). The former are usually associated with performance data and they include two subcategories: Channel Voice and Channel Mode. System messages usually refer to data that address the entire MIDI system and are divided in three subcategories: System Common, Real Time, and Exclusive. For the scope of this book the one we really need to understand is the Channel Voice group. These data are the ones used the most when dealing with software synthesizers and modern studio setups.

Structure of MIDI data

A typical MIDI data is formed by two or more bytes (remember that each byte contains 8 bits). The first byte of a MIDI message is called the Status byte, and the following bytes are called Data bytes. Remember that each byte is sent one at the time because MIDI is a serial protocol. The role of the Status byte is to send two very important pieces of information: the type of message sent and the MIDI channel to which the message is sent. In the MIDI standard there are 16 channels available (from 0 to 15). The structure of a typical Status byte for a Note On message can be seen in Figure 3.59.

As you can see in Figure 3.59 the first bit of the Status byte is reserved as identifier for the type of byte (Status bytes begins with 1, and Data bytes begin with 0). The next three bits are reserved for the type of data sent. For example, a Note On message is described as 001, or a Note Off message is represented by 000. The last four bites of a Status byte describe the MIDI channel on which the message is sent. They range from 0000 (channel 1) to 1111 (channel 16). At this point we need to dig just a bit deeper into how the binary system works in order to better understand the MIDI data structure.

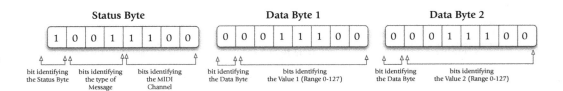

Figure 3.59
The Status byte for a Note On message, sent on channel 1.

A Binary World

The binary system is a way of describing any number by using only two digits: 1 and 0 (bits). This system is used by computers and electronic circuits because 1 and 0 can be easily represented by "on" and "off" status in a circuit. Now, you might ask, "how is it possible to represent any number using only two digits?" The answer is simple: by concatenating several bits together and assigning an exponentially increasing value to the next bit. This may sound complicated, but, in fact, it is not. If we have a system that has only one bit available we can represent only two numbers: 0 and 1 (Figure 3.60).

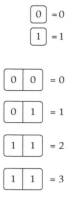

Figure 3.60
A simple "1-bit" binary system.

Figure 3.61
A "2-bit" binary system.

If, to this system a second bit is added, now we have four different combinations available: 00, 01, 10, and 11. Therefore we can represent four numbers in total, respectively: 0, 1, 2, and 3 (Figure 3.61). Each additional bit in a binary system allows us to double the numbers we can represent with that system.

It is very simple to find out how many numbers can be described by a certain bit system. The formula is 2 power of n, where n is the number of bit of the system. For example a 4-bit system allows me to represent sixteen numbers (from 0 to 15).

The MIDI Channels

As we mentioned earlier, the channel on which a MIDI message is sent is specified in the last four bits of the Status byte. As we just learned, with a 4-bit system we can describe sixteen values, that's why the MIDI standard is based on sixteen channels. Keep in mind that in a binary system number 0 is as important as any other number. Therefore in the "MIDI world" MIDI channels range from 0 to 15. In the "real world," though, this would be a bit inconvenient, and therefore we refer to the MIDI channels as ranging from 1 to 16, even though they are really sent in binary format as 0–15.

Type of MIDI Messages

Now we should be able to assemble a complete Status byte. Let's try. The only piece missing is the table with all the most important Channel Voice messages and their respective binary codes. Let's take a look at Table 3.7.

Table 3.7 Channel Voice Messages With Their Status Bytes and Data Bytes

Status	Message Description	Data Bytes
1000	Note Off	2 (pitch, velocity)
1001	Note On	2 (pitch, velocity)
1010	Polyphonic Aftertouch	2 (pitch, pressure)
1011	Control Change	2 (id, value)
1100	Program Change	1 (program)
1101	Channel Aftertouch	1 (pressure)
1110	Pitch Bend	2 (LSB, MSB)

In Table 3.7 you can see how the first four bits of the Status byte start always with 1 and then are followed by three additional bytes that describe the type of messages sent. So, to send a Note On on channel 1 we will have the following Status byte: 1001 (Note On) + 0000 (MIDI channel 1). Let's try another one. If we want to send a Control Change message on MIDI channel 6 we will have 1011 (Control Change) + 0101 (MIDI channel 6).

Data Bytes

The second part of a Channel Voice messages includes the so-called Data bytes. In these bytes we send information that describes the details of the messages specified in the Status byte. As you can see in Table 3.7, each type of message has one or more data bytes that are specific to that particular message. For example, a Note On message needs two additional Data bytes, one for the note number (which describes which note we pressed) and another one for the velocity of the note (how hard we pressed that key on our MIDI controller).

It is important to remember that every Data byte starts always with a 0; this is to differentiate between a Status byte (which always starts with a 1) and a Data byte. Because a Data byte starts always with a 0 we really have only seven bits available in each Data bytes to describe the parameter of any given MIDI message. This is the reason why the majority of the parameters of Channel Voice messages have a range of 0–127 (remember that the formula used to find the total number of steps of a given bit system is 2 power of *n*; therefore 2 power of 7 is 128).

Now let's try to assemble a complete MIDI message. If we want to send, for example, a Note On message, on channel 3, with a note number 60 (middle C), and velocity 64 the message would look like this:

Status byte: 1001 (Note On) + 0010 (MIDI channel 3)
1 Data byte: 00111100 (note number 60)
2 Data byte: 01000000 (velocity 64)

Keep in mind that some channel voice messages can have parameters that have a higher resolution than 128 steps. The Pitch Bend, for example, has a range of 16,383 steps. This is achieved by combining two Data bytes together (one called the Most-Significant byte or MSB, and the other called the Least-Significant byte or LSB). By combining two bytes together we practically have now fourteen bits available (not sixteen because remember that the first bit of each Data byte needs always to be set to 0).

Description of MIDI Messages

Let's take a closer look at each channel voice message in order to better understand what it does.

Note On message: This message is sent every time you press a key on a MIDI controller; as soon as you press it, a MIDI message (in binary code) is sent to the MIDI OUT of the transmitting device. The Note On message includes information about the note pressed (C-2 to G-8 with values between 0 and 127), the MIDI channel (1 to 16), and the velocity-on parameter (0–127), describing how hard you press the key.

Note Off message: This is sent when you release the key of the controller. It terminates the note triggered with a Note On message. You can achieve the same result by sending a Note On message with its velocity set to 0, which helps reduce the stream of MIDI data to the network. It also contains the velocity-off parameter, which registers how hard you released the key (note that this particular information is rarely used).

Aftertouch: This is a MIDI message sent after the Note On message. If you push a little bit harder on the key of the keyboard controller, after the first hit, a MIDI message called Aftertouch is sent to the MIDI OUT port. The Aftertouch message can be assigned to control the vibrato effect of a sound, but it can also affect other parameters, such as volume, filter, and more. The Aftertouch can be polyphonic or monophonic. Monophonic Aftertouch (the most common type) is applied to the entire range of the keyboard independently of the key (or keys) that triggered it. Polyphonic Aftertouch, on the other end, allows you to send an independent message for each key, making it a more flexible option.

Pitch Bend: This message is controlled by the Pitch Bend wheel on a keyboard controller. It raises or lowers the pitch of the notes being played. It is one of the few MIDI data that has a range above the standard 128 steps. The range of this MIDI message extends from 0 to 16,383. Commonly, a sequencer would display 0 as the center position (non-transposed), 8191 fully raised, and –8192 fully lowered.

Program Change: This message is used to change the patch assigned to a certain MIDI channel. Each synthesizer has a series of programs (patches, presets, instruments, or sounds) stored in its internal memory. To each MIDI channel we can assign a specific patch by sending a program change message from a controller or a sequencer. The range of this message is 0 to 127. Because nowadays programs are organized into banks, where each bank stores a maximum of 128 patches, in order to change a patch through MIDI messages it is necessary to combine a Bank Change message and a program change message. Most devices use CC 0 or CC 32 (or sometimes a combination of both) to change bank.

Control Changes (CCs): These messages allow you to control specific parameters of a MIDI channel. There are a total of 128 CCs (0–127); the range of each controller ranges from 0 to 127. Some of the most standard controllers include CCs 1, 7, 10, 11, and 64. CC 1 is assigned to Modulation. It is sent through the Modulation wheel on a keyboard controller. Although in the past it was associated with a slow vibrato effect, nowadays it can be assigned to control pretty much any parameter of a software synthesizer. CC 7 controls the volume of a MIDI channel from 0 to 127, and CC 10 controls its stereo image positioning (pan). Value 0 is pan hard left, 127 is hard right, and 64 is centered. Controller 11 is assigned to Expression; it is similar to volume but it represents a percentage of CC 7. Controller 64 is assigned to the Sustain pedal (the notes played are held until the pedal is released). This controller has only two positions: on (values 64) and off (values 63). Table 3.8 lists all 128 controllers with their specifications and their most common uses in sequencing situations.

Table 3.8 Channel MIDI CC Messages

Controller #	Function	Usage
0	Bank select	Allows you to switch bank for patch selection. It is sometimes used in conjunction with CC#32 in order to send bank number higher than 128.
1	Modulation	Sets the modulation wheel to the specified value. Usually this parameter controls a vibrato effect generated through a LFO. It can also be used to control other sound parameters such as volume in certain sound libraries.
2	Breath controller	Can be set to affect several parameters but usually is associated with aftertouch messages.
3	Undefined	
4	foot controller	Can be set to affect several parameters but usually is associated with aftertouch messages.
5	*Portamento* value	Controls the rate used by *portamento* to slide between two subsequent notes.
6	Data entry (MSB)	Controls the value of either registered parameter number (RPN) or nonregistered parameter number (NRPN) parameters).
7	Volume	Controls the volume level of a MIDI channel.
8	Balance	Controls the balance (left and right) of a MIDI channel. It is mostly used on patches that contain stereo elements (such as stereo patches). 64 = center, 127= 100% right, and 0 =100% left.
9	Undefined	
10	pan	Controls the pan of a MIDI channel. 64 = center, 127 = 100% right, and 0=100% left
11	Expression	Controls a percentage of volume (CC#7).
12	Effect controller 1	Mostly used to control the effect parameter of one of the internal effects of a synthesizer (for example, the decay time of a reverb).
13	Effect controller 2	Mostly used to control the effect parameter of one of the internal effects of a synthesizer.
14–15	Undefined	
16–19	general purpose	These controllers are open and they can be assigned to aftertouch or similar messages.
20–31	Undefined	
32–63	LSB for Controls 0–31	These controllers allow you to have a "finer" scale for the corresponding controllers 0–31.
64	Sustain pedal	Controls the sustain function of a MIDI channel. It has only two positions: off (values between 0 and 63) and on (values between 64 and 127).

Table 3.8 Continued

Controller #	Function	Usage
65	*Portamento* on/off	Controls if the *portamento* effect (slide between two consequent notes) is on or off. It has only two positions: off (values between 0 and 63) and on (values between 64 and 127).
66	*Sostenuto* On/Off	Similar to the sustain controller, but holds only the notes that are already turned on when the pedal was pressed. It is ideal for the "chord hold" function, where you can have one chord holding while playing a melody on top. It has only two positions: off (values between 0 and 63) and on (values between 64 and 127).
67	Soft pedal on/off	Lowers the volume of the notes that are played. It has only two positions: off (values between 0 and 63) and on (values between 64 and 127).
68	*Legato* footswitch	Produces a *legato* effect (two subsequent notes without pause in between). It has only two positions: off (values between 0 and 63) and on (values between 64 and 127).
69	Hold 2	Prolongs the release of the note (or notes) playing while the controller is on. Unlike the sustain controller (CC#64), the notes won't sustain until you release the pedal but instead they will fade out according to their release parameter.
70	Sound controller 1	Usually associated with the way the synthesizer produces the sound. It can control, for example, the sample rate of a waveform in a wavetable synthesizer.
71	Sound controller 2	Controls the envelope over time of the VCF of a sound, allowing you to change over time the shape of the filter. It is also refered to as "resonance."
72	Sound controller 3	Controls the release stage of the VCA of a sound, allowing you to adjust the sustain time of each note.
73	Sound controller 4	Controls the attack stage of the VCA of a sound, allowing you to adjust the time that the waveform takes to reach its maximum amplitude.
74	Sound controller 5	Controls the filter cutoff frequency of the VCF, allowing you to change the brightness of the sound.

(continued)

Table 3.8 Continued

Controller #	Function	Usage
75–79	Sound controllers 6–10	Generic controllers that can be assigned by a manufacturer to control nonstandard parameters of a sound generator.
80–83	General-purpose controllers	Generic button-switch controllers that can be assigned to various on/off parameters. They have only two positions: off (values between 0 and 63) and on (values between 64 and 127).
84	*Portamento* control	Controls the amount of *portamento*.
85–90	Undefined	
91	Effect 1 depth	Controls the depth of effect 1 (mostly used to control the reverb send amount).
92	Effect 2 depth	Controls the depth of effect 2 (mostly used to control the tremolo amount).
93	Effect 3 depth	Controls the depth of effect 3 (mostly used to control the chorus amount).
94	Effect 4 depth	Controls the depth of effect 4 (mostly used to control the celeste or detune amount).
95	Effect 5 depth	Controls the depth of effect 5 (mostly used to control the phaser effect amount).
96	Data increment (+1)	Mainly used to send an increment of data for RPN and NRPN messages.
97	Data increment (−1)	Mainly used to send a decrement of data for RPN and NRPN messages.
98	NRPN LSB	Selects the NRPN parameter targeted by controllers 6, 38, 96, and 97.
99	NRPN MSB	Selects the NRPN parameter targeted by controllers 6, 38, 96 and 97.
100	RPN LSB	Selects the RPN parameter targeted by controllers 6, 38, 96, and 97.
101	RPN MSB	Selects the RPN parameter targeted by controllers 6, 38, 96, and 97.
102–119	Undefined	
120	All sound off	Mutes all sounding notes regardless of their release time and regardless of whether the sustain pedal is pressed.
121	Reset all controllers	Resets all the controllers to their default status.
122	Local on/off	Enables you to turn the internal connection between the keyboard and its sound generator on or off. If you use your MIDI synthesizer on a MIDI network, most likely you will need the local to be turned off in order to avoid notes being played twice.

Table 3.8 Continued

Controller #	Function	Usage
123	All notes off	Mutes all sounding notes. The notes that are turned off by this message will still retain their natural release time. Notes that are held by a sustain pedal will not be turned off until the pedal is released.
124	Omni mode off	Sets the device to Omni off mode.
125	Omni mode on	Sets the device to Omni on mode.
126	Mono mode	Switches the device to monophonic operation.
127	Poly mode	Switches the device to polyphonic operation.

MIDI Connections

Every device that needs to be connected to a MIDI studio, or system, needs to have a MIDI interface. The MIDI standard uses three ports to control the data flow: IN, OUT, and THRU. The connectors for the three ports are all the same: a 5-pin-DIN female port on the device (Figure 3.62) and a corresponding male connector on the cable.

The OUT port sends out MIDI data generated from a device, and the IN port receives the data. The THRU port is used to send out an exact copy of the messages received from the IN port. Of course, as we mentioned earlier, a device, in order to be connected to the MIDI network, must be equipped with a MIDI interface. All professional electronic music instruments nowadays feature a built-in MIDI interface. In the modern production studio the presence of hardware gears is less prominent than it used to be in the past. The usage of software synthesizers has simplified the studio setup immensely. Of course, the computer in your studio running the DAW also needs to be equipped with a MIDI interface (usually a USB-connected external device).

Beyond MIDI: Open Sound Control

Although MIDI is still a solid protocol to connect different music devices together, new systems and protocols have been developed since 1982 in order to keep up with the latest technologies and provide musicians and producers with more powerful tools. The majority of these new systems are based on computer networking protocols such as Tcp or UDP. Although some of these systems (such as ipMIDI and MIDIoverLAN) have the goal

Figure 3.62
Standard MIDI ports (Courtesy of Roland Corporation US).

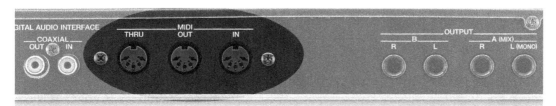

of simply transferring MIDI data over local area network (LAN) infrastructures, others have been trying to reinvent the way musical (but not only) devices can communicate with each other. This is the case of Open Sound Control (OSC). OSC is a networking protocol that allows musical instruments, computers, and other devices to communicate and exchange data. Its advantage over MIDI is that it is built over standard network protocols, therefore allowing for higher resolution and flexibility. OSC data can be transferred wirelessly as well over wired connections, not only over a LAN but also via the Internet. OSC has been implemented not only for simple musical data transfer but also for real-time sound and media processing, software synthesizers, programming languages, and real-time interactive projects.

The Electronic Digital Studio for the Contemporary Composer and Producer

Being able to work in a well-planned, organized, and efficient environment is crucial for the modern producer. When designing your work space there are two things you want to avoid: unnecessary expenses and structures that are nonergonomic. We strongly advise you to find a nice room with good light and ventilation. You will be working for many hours in this environment, so make yourself comfortable! The equipment that you will be using for you productions can be categorized into six main areas: the main computer (and the DAW running on it), the audio interface, the controller(s), monitoring equipment (speakers), recording equipment (microphones and preamplifiers), and software (or hardware) synthesizers. Let's take a look at these categories.

The Computer

The computer is at the center of your production studio. It serves not only as the "central dispatcher" of all your MIDI and audio signals, but also as the machine on which your software synthesizers run. Although it is not in the scope of this book to go into the details of which type and model of computer you should get, here are few key tips. The basic idea, when selecting a machine for music production, is that you want the fastest and most powerful computer you can afford. Modern software synthesizers and plug-ins can require a considerable amount of CPU power in order to run in real time an incredible number of calculations. This is particularly true for types of synthesis such as Additive or Granular. Therefore try to get the most powerful machine possible. This translates in selecting a computer that has the highest CPU speed (measured in gigahertz) and number of cores (these days four cores is the bare minimum we would recommend). Another factor that has a big impact on the overall speed of your machines is the amount of RAM, which is measured in gigabytes. For a power user 16 GB of RAM is the minimum we would recommend, and 32 GB would be a better choice if your budget allows it. Having plenty of storage space is also very important, particularly if you are planning to use large sampling libraries (such as orchestral libraries). Choose fast HDs (at least 7200 RPM) with sizes of 1 TB or higher. Even better is the route of using solid state drives (SSDs), which are much faster and reliable. If your budget doesn't allow you to get all SSDs, then, at least, use a SSD for the operating system. This will improve the overall performance of your computer exponentially. Having plenty of expansion ports (internally and externally) will guarantee that your machine will stand the passing of time in a more robust way. We recommend having at least five USB 3 ports, 1 or 2 e-SATA, 6 SATA, and 2 Thunderbolt ports. Nowadays having two screens is a must for any serious music production or sound design. Two 23" or 27" monitors will give a nice work surface.

The Audio Interface

The audio interface is responsible of the digital-to-analog conversion (DAC) and analog-to-digital (ADC) conversion of the audio signal generated and recorded, respectively, by your DAW and software synthesizers. This is a crucial component in regard to audio quality for both recording and playback. Although we will get into the details of ADC/DAC later, understanding the main feature of an audio interface will help you in deciding which type and model to get for your studio. First, you need to decide the type of connection(s) your interface will use to connect to your computer. Audio interfaces feature mainly four different types of connectors: PCI, USB, Thunderbolt, and the older FireWire (replaced by Thunderbolt). For some pros and cons of each type of connection take a look at Table 3.9.

The majority of the interfaces feature only one type of connector but some (particularly the newest ones) may have two types, such as Thunderbolt and USB. Although up to a few years ago the type of connection to the computer used to be a big issue, nowadays all the types listed in Table 3.9 provide enough bandwidth for any audio need.

The other three main components of the audio interface to consider are the converter, the number of I/Os, and the number of microphone preamps available. The quality of the converters (ADC/DAC) is crucial to achieve a good sound. You need an interface that is able to record at up to 96 kHz and 24 bits in order to obtain professional results. Regarding the number of I/Os, usually the more the better, but this depends on how many individual tracks you are planning to record simultaneously. For the majority of electronic music–based projects we think that two INs and two OUTs are sufficient. If you are planning to work on surround projects though you need at least six outputs in order to feed the five satellite speakers and the subwoofer. In general, for electronic projects we would rather have fewer inputs but a higher end interface with better converters. The number of built-in microphone preamplifiers depends on how many simultaneously tracks you plan to record with microphones. Two is really the minimum (for stereo recordings), but four would give you a bit more flexibility. We prefer an audio interface with two (or four) built-in preamps and other two or more line INs to which, if necessary, we can add extra external third-party preamps such as the 500 series modules for the API lunchbox (Figure 3.63).

The Controller

For your electronic production studio you definitely need one or more MIDI controller. As we learned earlier, a MIDI keyboard controller features a piano keyboard and a MIDI interface, but usually no internal built-in sound generator. Having a good keyboard controller is essential. Keyboard controllers come in different sizes. For electronic music production usually you don't need a full 88-key keyboard, A 49- or 61-key

Table 3.9 Types of Connections for Audio Interfaces

Type of Connection	Pros	Cons	Comments
PCI	• One of the fastest connections • Reliable • High bandwidth	• Available only for desktop computers because the connection resides on the motherboard, unless you use an expansion chassis • Easier to become obsolete by newer PCI standards • Requires extra steps for installation and removal	• Usually used for high-end systems like the AVID HDX
USB	• Very versatile • Cheap connection to implement • Great for portable systems ans desktops alike • Universal • Used by most interfaces that are "class compliant"	• Not as fast as other protocols but USB 3 (and its new iteration USB-C) has plenty of speed for audio applications	• USB is one of the most used connections, delivering affordable and versatile connections for basic as well as professional interfaces
FireWire	• Fast and reliable • Many of the pro interfaces use it	• A bit more expensive to implement • This protocol is being replaced with Thunderbolt • Two different types (FW400 and FW800) can be confusing	• FW has been the choice for many professionals when it comes to audio. It is being replaced with Thunderbolt
Thunderbolt	• One of the fastest interfaces available at the moment. • Newest standard	• Not very diffused yet • Expensive to include with motherboard • Cables are expensive • Limitation of seven devices on a chain	• This is the newest connection developed by Apple and Intel. There are three types called 1, 2 ,and 3. The connectors between the three types are the same but the speed increases for each new revision.

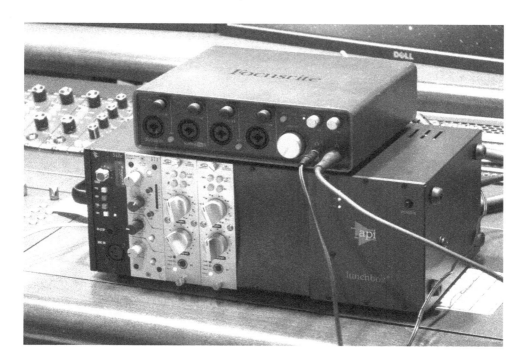

Figure 3.63
Audio interface connected to an external API Lunchbox with four extra preamps.

would be enough in most situations. We recommend getting the largest one you can afford. You will need the extra keys for patches that utilize key switches. There are also three main types of keyboard action you can choose from: weighted, semiweighted, and synth. Weighted controllers have the same response as (or at least a one very similar to) an acoustic piano. They are usually more expensive and much heavier than keyboards of synthesizers with plastic keys. If you are on a tight budget and the real piano feel is not a major concern, we suggest opting for a controller with a lighter key action. Semiweighted keyboard controllers have a medium action response. Their keys feature a solid action but definitely lighter than the fully weighted one. They have the advantage of being very versatile, offering a good response and also allowing for good performance speed. If you are not concerned with a realistic piano feel, then a controller with a synth light keyboard is probably a good choice. In fact, for electronic sequencing and production, this is the type that we would recommend. Make sure that your keyboard controller has plenty of assignable faders, knobs, and pads; they will come in handy when you are controlling the different parameters of your software synthesizers (Figure 3.64).

Figure 3.64
A MIDI keyboard controller with a great selection of assignable faders, knobs, and pads (courtesy of Novation).

Figure 3.65
A guitar-to-MIDI converter system.

These extra controllers are easily assignable to any MIDI control change. For example, you will be able to assign the speed of an LFO to one of the faders and control the cutoff frequency of a filter with another fader. The majority of the keyboard controllers connect to your computer via USB. Usually they also have a regular MIDI 5-pin DIN connector for interfacing directly to other MIDI devices or to a computer MIDI interface.

Of course there are other MIDI controllers that are not keyboard based. The most common alternative controllers are the guitar-to-MIDI controller (GMC) and the MIDI drum pad. Both are not only ideal respectively for guitarists and drummers, but also for producers who want to experiment with different systems of controlling their synthesizers. A guitar-to-MIDI controller (or guitar MIDI controller) allows a regular acoustic or electric guitar to be connected to a MIDI system. It outputs MIDI messages and notes to any MIDI device, including a DAW. The principle on which this type of controller is based is simple: a pickup (divided into six segments, one for each string) is mounted next to the bridge of the guitar (Figure 3.65).

The pickup detects the frequency of the notes played on each single string by analyzing their cycles; this information is passed to a breakout unit that converts the frequencies in MIDI note On and Off messages. From the unit, a regular MIDI OUT port sends the messages to the connected device(s). The pickup can also detect bending of the strings that is translated in pitch bend messages. This one is an alternative controller that is worth having in your studio as either your main controller or as an occasional substitute for your keyboard. As the price of these devices keeps getting lower we definitely recommend investing in one. A GMC can be used in creative ways. Because each string has an individual pickup, by assigning different strings to different MIDI channels and patches, you can create a "one-man band" and generate interesting layers and ensemble combinations. You can record-enable multiple tracks (maximum of six, one per string), all receiving on different MIDI channels, and assign their outputs to separate devices and MIDI channels. You might assign a low sustained pad to the low E and A strings, use the D and G strings for sound effects or tenor voice counterpoint, and the B and high E for the melody. The advantage of this approach is that you can capture a much better groove and "live feel" than sequencing the tracks separately.

MIDI pads are controllers meant to complement either a drum kit or a keyboard controller. They usually feature four, six, or eight pads that can be played with either sticks or hands. The MIDI channel and note assignment can be set for each pad individually. A creative way to program a percussion/drum controller is to use it as a MIDI "pitched" instrument by assigning different pads to different notes of a pitched instrument or synth patch. You can also assign certain pads to trigger loops or samples and use them as background layers for a one-man-band type of performance. No matter which one you choose, we highly recommend having one in your studio. You are not necessarily required to be a proficient drummer to use a MIDI pad. In fact, these devices are particularly appealing to nondrummers because you can take advantage of all the editing features of your sequencer to fix and improve your parts.

Figure 3.66
Breath controller USB MIDI Breath and Bite Controller 2 by TEControl (http://www.tecontrol.se).

Another type of MIDI controller that can be extremely useful and fun to play with is the breath controller (BC) (Figure 3.66).

This MIDI device connects to a standard USB port on your computer. It allows you to send any CC data by blowing air into its mouthpiece. You can program the type of CC sent through the BC through its dedicated software editor. With the USB MIDI Breath and Bite Controller 2 by TEControl (http://www.tecontrol.se), you can send MIDI CC messages not only with breath pressure but also with byte, and up/down/left/right head movements. You can assign the BC to CC#11 (Expression) and smoothly control the volume of your parts while playing. Or, assign it to CC#1 (Modulation), and you will be able to add a touch of vibrato. The applications are endless, and you will find yourself wondering how you could have sequenced before without it. If you are a wind instrument player we also recommend using a wind controller, such as the Akai EWI or the Yamaha WX series. The main advantage of these devices is that they provide a way to input notes that is very close to the experience of playing a real wind instrument. Their clarinet-like shape and controls allow you to sequence any type of lead part with extreme flexibility and expressivity. Some of the features that make these devices incredible sequencing tools are the ability to switch octaves (extending the range of the instrument), to assign one or more control wheels to arbitrary MIDI CC, and to select among several fingering options, such as clarinet, saxophone, and flute. The biggest drawback of these controllers is that they require you to be a very skillful musician in order to master all their sophisticated capabilities. Even the most experienced veteran wind instrument player will need a good amount of time to adjust to the differences between an acoustic instrument and the controller. If your main instrument is not saxophone, clarinet, or flute, we suggest using a keyboard controller in combination with the BC.

Monitoring System

We can't stress enough how important is to have a good monitoring system in your studio. A good set of speakers can go a long way in helping you achieve great sonic results. You can opt for passive speakers combined with a separate amplifier or for active speakers that have built-in amplifiers. We recommend the second option because usually the amplifier in active speakers is calibrated to give you the best sound for those particular monitors. Depending on the design, use, and distance at which they provide the best performance, speakers can be divided into *near-field*, *mid-field*, and *far-field* types. In a project studio you usually are going to use near-field or mid-field monitors, which are designed for mixing at a short or medium distance, respectively, to avoid the disturbances introduced by a faulty design of the room. For detailed information on audio monitors and speakers we recommend the comprehensive text by John Borwick, *Loudspeaker and Headphone Handbook*, published by Focal Press. We suggest, if your budget allows it, to get speakers that are able to autocorrect their frequency response to the room in which they are placed. Usually they come with a special microphone and software that allows you to measure the frequency response of your room (and to detect the extra coloration of the listening environment). Once the sonic "imprint" of the room is taken, the software will be able to apply the necessary equalization in order to get a response as flat as possible.

Microphones and Preamplifiers

Any studio or project studio needs to be equipped with good microphones and mic-preamplifiers. Even if you plan to concentrate on electronic production, there will be plenty of situations in which you will need to record live sources, be it for sampling some new material that will be processed and synthesized at a later stage or to record live instruments for hybrid productions. The concept of a preamplifier is quite simple. The waveform captured by the diaphragm of a microphone is transformed by a transducer into electric signal. This signal is "weak," in the range of 1–100 mV. To interface with other components of your studio the signal needs to be amplified to a line level (up to 10 V). This is where a mic preamplifier comes in. Usually your audio interface comes equipped with built-in preamps (two or four preamps are usually the norm). Depending on the type and price of the interface, the quality of the built-in preamps can range from just acceptable all the way to excellent. An entry-level interface features, usually, preamps that can be noisy (low signal-to-noise ratio) and can color the sound in undesirable ways. A top-notch interface, on the other end, can be equipped with professional preamps that can produce detailed and transparent sounds. For an interface that has less-than-ideal preamps, fortunately, there is a solution. You can easily buy additional external preamplifiers, not only

Table 3.10 Comparison Among Different Types of Microphone Preamplifiers

Type of Preamplifier	Description	Sonic Qualities
Solid state	These preamps use transistor or op-amps to increase the signal's gain. They are very efficient and produce little heat.	• Clean and transparent • Low distortion even at high levels • Very accurate
Tube	These preamps use thermionic tubes (or valves) to obtain gain's increase	• Fatter with more color • Add more character and warmth to your original sound • Can add a nice controlled distortion if needed • Natural "built" compression as gain increases
Hybrid	These preamps combine both solid state and tube components. The transistors are used at the input stage and the tubes are used at the output.	• Clear and defined with the possibility of adding additional color and/or saturation
Transformer-less	These preamps don't have a transformer in the signal path between input and output. Solid state preamps are most common in the transformerless type	• The lack of the transformer makes the sound clear and transparent, particularly in the high frequencies
Transformer-coupled	The transformers bridge the input to the output while increasing the signal voltage. They offer better circuit protection and lower levels of noise.	• Quieter with less "hum" noise

to improve on the interface's ones, but also to expand your selection. Preamps have different signature sounds that depend on the manufacturer, the circuitry, and the type of construction. We suggest having at least three or four different types of preamp in your studio to be able to pair different types of microphones with different types of sources, and ultimately be able to create your signature sounds. For a quick comparison among different types of preamplifiers take a look at Table 3.10.

External preamps connect to your audio system via the built-in line inputs of your audio interface. The signal flow is therefore microphone, mic in of the external preamp, line out of the external preamp, and line in of the audio interface. This is why it is important to have an interface that has also two or four line inputs. They will definitely come in handy when trying to expand the preamp selection of our studio. One of my favorite solutions to easily have different types of preamps available is the API 500 format (see Figure 3.63).

Table 3.11 Main Types of Microphones

Microphone Type	Characteristics	Sonic Features
Dynamic	• Most dynamic microphones do not go above 16 kHz • Great for live performance (very sturdy) • Good for loud sources that are close to the mic	• Usually sound darker than condenser mics • Tend to also feel a bit light on the bass range • Usually have only cardioid pattern
Condenser	• Full frequency spectrum • Great definition • Mainly used in studio recording • Different patterns available	• Can capture a full and bright sound yet with a warm characteristic
Ribbon	• Very delicate • Usually in figure-eight pattern	• Their main sonic feature is warm and smooth

This unit provides power and I/Os to a series of empty slots that can be filled with units from different manufacturers. Each module is less expensive than its stand-alone counterpart because it doesn't have to include the circuitry for power and I/Os. In addition the system is very compact, making it ideal for small studios or for portable rigs.

Even if you don't plan to record live ensembles or multitrack live instruments, having a decent selection of versatile microphones is very important. In general we recommend at least five or six good microphones with different characteristics and patterns. This will provide a good balance between different colors, giving you great versatility. When considering a microphone you need to look at three main features: type, pattern, and size of the diaphragm.

Microphone Types

The three main types of microphone designs are dynamic, condenser, and ribbon. As it was the case for the preamps, each type has different applications and sonic characteristics. Look at Table 3.11 for a description of each type.

Synthesizers

The category of "synthesizers" basically covers everything that you will need and use to create your music and work on your productions. The rest of this book is dedicated to this category. We will methodically go through the most important types of synthesis and synthesizers so that you can have a complete view of what to have in your studio and how to use it to create interesting sounds and productions. The types of synthesis that you should be familiar with and that you should have available in your studio areas follows:

Subtractive
Frequency Modulation

Table 3.12 Studio Equipment Options

Equipment	Home/Project	Portable	Hybrid	Comments
Computer	• Full desktop with large amount of RAM (32 GB)	• Light laptop or iPad	• Powerful laptop with large amount of RAM (16 or 32 GB)	• For a fully portable studio you want to have a light laptop that allows you to travel light. For a hybrid option you need to go for a powerful laptop that allows you to work in your home studio too.
Audio interface	• Great quality with great converters and mic preamps • You can use PCI, Thunderbolt, or any other type of connection	• A light and well built audio interface. • USBs or Thunderbolts are preferred connectors	• A light and well-built audio interface. • USBs or Thunderbolts are preferred connectors • Good converters and two built-in mic–preamps	
Controller	• Full 88 keys with assignably faders and sliders	• 25 or 49 keys with synth action (not weighted) for extreme portability	• 25 or 49 keys for traveling and 61 or 88 keys at home for more serious studio work	
Monitoring system	• Set of amplified studio monitors	• Set of good headphones	• Amplified monitors and headphones	
Recording equipment (mics & preamps)	• Good selections of preamplifiers and microhpnpes	• For mic– preamps and API Lunchbox (two or four spaces) with some nice high-end preamplifiers • One or two large-diaphragm condenser microphones • Two matched small-diaphragm condenser microphones for stereo recordings	• For mic preamps and API Lunchbox (six spaces) with some nice high-end preamplifiers • Two large-diaphragm condenser microphones • Two matched small- diaphragm condenser microphones for stereo recordings • One or two ribbon microphones	
Software	• At least two DAWs • Great selection of software synthesizers • Professional sample libraries	• One DAW • Good selection of major software synthesizers • Light version of sample libraries	• At least two DAWs • Great selection of software synthesizers • Professional sample libraries	• For the hybrid option have some external HDs with the full libraries/soft synths and some portable ones with lighter (smaller) versions of the same libraries for when you travel

Additive
Physical Modeling
Wavetable
Sample-based
Granular

Portable vs. Home Studios

All the elements that we described so far that constitute your studio can, generally speaking, be adapted for either a home/project studio or a portable setup. The choice of setup really depends on your needs and your work habits. Usually we recommend having both, a home studio where you can mix, record live instruments, and receive clients for final approvals, and a portable for sketching and preproduction stages for when you are traveling. If you are not ready to commit to a portable and a home setup right away, then you need to pick one. In order to make the correct decision here are some questions that you should ask yourself first:

Out of 365, how many days a year are you traveling or away from your home?
How much music writing/production are you asked to do when you are away?
How much music writing/production are you asked to do with very short deadlines?
How much computer power do you really need?
How much audio recording you do for your productions?
Do you plan to play live with your setup?

Answering these basic questions will help you determine which setup you should consider first. For each studio (home or portable) the categories that we discussed previously still apply. You will still need a computer, a monitoring system, a control, and so forth. The size and features of these items though will be different depending on the setup. Let's take a look at Table 3.12 to compare the two approaches.

As you notice, in Table 3.12, we added a third option called "Hybrid"; this is the option for producers who are split between studio and "on the road" work. In this case you need a powerful laptop capable of also handling heavy studio work. Some of your gear will need to be duplicated for the studio and traveling setups like, for example, the keyboard controller. In this case you will need a portable (twenty-five or forty-nine keys) one for traveling and a studio one (eighty-eight keys) for studio work.

Now that you have mastered all the necessary audio and synthesis concepts and the main components of your studio, it is time to dive into making some sounds and music!

4

Subtractive Synthesis

Introduction

Time to get making sounds! Starting with this chapter and throughout the remainder of the book, chapters are formatted in a way that starts with background and foundational information followed by a section called "Producer Point of View" in which specifics on how sound designers can use the corresponding chapter's technique in their work.

Sculpting Sound With Subtractive Synthesis

Whether starting from a block of stone or ice or a slab of wood, a sculptor envisions an embedded form within that will be exposed over time by chiseling away the material that conceals it (see Figure 4.1). This is an apt analogy to the process the sound designer employs when working in subtractive synthesis. The raw block of material is a timbre rich in overtones like a sawtooth or square wave or noise. In place of a hammer and chisel, the sound designer uses filters and effects. Unlike the finished sculpture, however, time is inherent in sound-based art forms; therefore, dynamically shifting the filters and effects with LFOs and EGs is important for adding variation and interest.

Subtractive synthesis is the first technique covered in this book because it was the first widely used form of synthesis. The RCA Electronic Music Synthesizer, the Theremin, the Moog Modular, and the Buchla 100 all used oscillators as sound generators and filters

Figure 4.1
William deGarthe's granite monument dedicated to local fisherman in Peggy's Cove, Nova Scotia, Canada.

Creating Sounds from Scratch

Figure 4.2
Flowchart of subtractive synth.

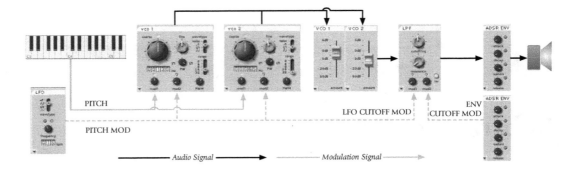

and other effects as modifiers. This configuration of generator–modifier is central to nearly all forms of synthesis that were to follow.

The "subtractive" nature of this technique necessitates beginning with a timbre that is complex and rich in overtones. The source of this sound is a standard waveform (triangle, square, pulse, and noise) generated by an oscillator or played back as a sample from a digital synthesizer. You may have noticed "sine" was omitted from the list of standard waveforms; it is indeed a standard waveform, but one that is of lesser use as a generator in subtractive synthesis because there is only a fundamental, not overtones, to chisel away!

A lot can be done with a single oscillator feeding into a modifier stage, and there are many examples of instruments that do this quite well, but working with two to three oscillators is common and preferable. Figure 4.2 diagrams a simple configuration using Applied Acoustics Systems' Tassman. Tassman is easy to use, extremely flexible, and a great-sounding synth that is a helpful place to develop an understanding of modular synthesis because it gives you lots of modules to choose from to build an instrument to your liking—plug and play, try things out, you can't break anything!

> *Note:* See this book's companion website for details on an evaluation copies and discounts on Applied Acoustics Systems (AAS) instruments. For what it's worth, we approached *them*—not the other way around—because we have a lot of respect for their products.

The flowchart of Figure 4.2 shows a keyboard as the controller. A conventional music keyboard using standard tuning has twelve semitones per octave. Commonly on analog synths, an octave up/down represents a change of ± 1 V, each semitone representing a change of 1/12th of a volt. In this scenario, the keyboard is acting as a modifier to the oscillator by controlling its frequency. The Tassman VCOs show a vertical switch for selecting a waveform: noise, sawtooth, square, and sine. Beneath that is a switch labeled "Range" for selecting the octave range in which the oscillator sits: 2, 4, 8, 16, or 32. The numbers correspond to lengths in feet of a pipe, like those found on a pipe organ on which the smaller lengths resonate at higher frequencies than the larger ones. Playing in octaves can easily be done by choosing a waveform other than noise and setting the ranges of two VCOs to 2 and 4, or 4 and 8, etc.

Notice in Figure 4.2 that each VCO has two modulator inputs: We are using mod1 for keyboard pitch. Turning mod1 to 12 o'clock accepts the incoming voltage signal from the keyboard as standard tuning—the scale can be stretched or compressed by boosting or attenuating above or below that point for unconventional tuning. The second modulation input—mod2—is being used for vibrato. When turned all the way down, the pitch is steady; as it is turned up, the vibrato increases in intensity. The speed and shape of the vibrato come from the wave options in the LFO module.

The output of each VCO is fed into a separate channel in the mixer for balancing. The mixer's output then goes through a LPF that has two modulation inputs to modulate the location of the cutoff frequency. In our design, mod1 connects the cutoff frequency with the LFO, cycling regularly over time. Mod2 is coming from an envelope generator so the timbre could, for example, start off brighter with the frequency set high and decay down to a lower sustain level, thus mimicking what acoustic instruments do naturally. The intensity of the cutoff shift is dependent on the level set in mod2. By and large, there is little variation between VCOs used by different manufacturers. Where the real sonic distinction becomes evident is in the design of the filters.

A second envelope here is used for volume. Its input is fed from the filter's output. Turning the sustain level all the way down isolates the attack and decay stages and will emulate a plucked or percussive sound that has no sustain. Turning the sustain level all the way up negates the decay stage because that is the time it takes to go from the maximum level down to the sustain level. Release is the time it takes for the sustain level to drop to zero. For that reason, some instruments will have an AD or AR envelope (rather than the standard ADSR) that assumes a sustain level of zero.

Conventional subtractive synthesis with traditional waveforms produces a sound that may seem a little dated. There are times, however, when those are the timbres needed on a given project or as a key layer of a more complex sound. Let's take a moment to explore some sound designs to get the sound of subtractive synthesis in your mind's ear, and make the process more familiar.

Included throughout the book are "sound designs" to help you better understand the topic under discussion. There is no need to use the exact instrument in our examples; there are many great instruments—both software and hardware—that can be used, and we made an effort to keep the instructions as generic as possible.

Fundamentals of Working With Subtractive Synthesis

Not all instruments that utilize subtractive synthesis are configured exactly the same, but there are certain components you can expect to find. The first step when approaching a new instrument is studying the available sources (oscillators in this case), the modifiers, and the routing capabilities.

As a case study, let's look at TAL-NoiseMaker, a free (donation requested) and very usable subtractive synth. There is not a lot hidden in this instrument so it a good one to start with. Take a look at Figure 4.3 first, and then we will go over the basics.

- Starting in the "Synth 1" section we see three oscillators: OSC 1 and 2, and one more lurking in the "Master" section labeled SUB. The Master section has three knobs for mixing levels from each of these three oscillators—overall volume and pitch control, ring modulation, oscillator sync, *portamento*, and unison control (labeled VOICES). OSC 1 and OSC 2 have pop-up menus for selecting sine, triangle, pulse, and sawtooth waves and noise; coarse tuning (TUNE) and FINE tuning; a knob for shifting the phase (starting position) of the selected wave; and when Pulse is the selected waveform, PW shifts the position of the duty cycle.

Creating Sounds from Scratch

Figure 4.3
Subtractive overview using TAL NoizeMaker.

- There are two LFOs with the usual wave and noise suspects as selectable patterns and a pop-up box next to AMOUNT where the output can be assigned (OSC 1, OSC 2, Filter, etc.)
- The FILTER section has the familiar ADSR controls with CUTOFF frequency, resonance amount, and a knob to control the cutoff frequency relative to the note being played (KEY), and the contour (CONT) or the shape of the paths connecting each stage of the envelope (linear to logarithmic).
- ADSR controls the main volume envelope with an extra AD-only envelope routable to destinations such as the filter, individual oscillators, PW, etc.

Lots of control for such a seemingly simple instrument! The compact, all-in-one design of an instrument like this is efficient—especially for hardware instruments—but can be a bit intimidating until you get a feel for what to expect. An instrument like Tassman is particularly useful for visualizing how audio and control signals are routed around an instrument. Once you develop that understanding, it becomes easier to look at an instrument with an interface of knobs and faders (or even simply text entry) and visualize what is going on "under the hood."

Let's make a sound and put these ideas into practice. We recommend downloading a demo of Tassman from our website or you can adapt the steps to any subtractive instrument you have available.

Ok, so it doesn't *really* sound like a clarinet, but the square wave has a distinct clarinet-like character, and it is a good one to get in your mind's ear. If you haven't done

Sound Design 4.1—Synth Clarinet

Steps (see Figure 4.4)

Figure 4.4
Sound Design 4.1 clarinet timbre using subtractive synthesis.

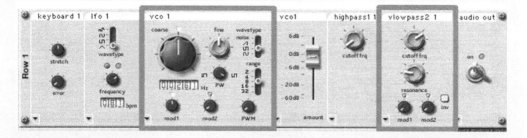

1. Select a subtractive-based instrument that has at least one VCO and a LPF.
2. Engage the "mono" switch on your synthesizer to ensure that only a single note is sounded at a time. In the case of Tassman, a mono keyboard was chosen over a "poly" (or polyphonic) keyboard.
3. Start with a VCO that has a pulse- or square-wave setting. In Tassman, the square-wave selection enables a range of pulse options (see "PWM" at the bottom of VCO 1). Notice that the control is turned completely in the clockwise direction—which, in the case of Tassman, represents a pure square wave.
4. Use a LPF to soften the higher harmonics by turning down the cutoff frequency to taste (meaning to where you think it sounds most like a clarinet) and experiment with adding a little resonance.

so already, go play with the controls beyond the recommended settings and see what happens. Bring up the resonance in the LPF, sweet the cutoff frequency, and so forth. If you push resonance far enough it may begin to self-oscillate—perhaps not ideal in this design, but a characteristic that can add a layer of interest, especially when you are working with a synth that has only one VCO.

Using an instrument with more than one VCO opens up a lot of possibilities, such as (1) layering different wave types for more complex timbres, (2) shifting the VCOs into separate octaves for big sounds, or (3) layering two sawtooth or pulse waves slightly detuned with each other to add depth.

A classic use of detuning is for creating synth strings. Nobody is likely to be fooled into thinking they are hearing a real string orchestra, but the timbre has become a real classic. The instrument favored by many musicians who did this so well in its day was Sequential Circuits' Prophet 5. Described in more detail in Chapter 1, it was the first truly polyphonic synth allowing chords to be played in realtime on the keyboard. Sound Design 4.2 uses Arturia's Prophet V model of the classic instrument to create a simple but effective synth string patch right out of the 1980s.

Sound Design 4.2—1980s Classic Synth Strings

In this example (see Figure 4.5) we make a basic synth strings patch that has a bright, harsh attack that mellows out quickly. We accomplish the mellowing out by lowering the filter's cutoff frequency over time with control from the envelope generator. In the following steps are the settings on Arturia's Prophet V emulation, but a similar sound can be achieved with any instrument that offers two oscillators with sawtooths and a LPF with envelope control.

Figure 4.5

Sound Design 4.2 1980s synth strings on Arturia's Prophet V.

Steps

1. Unlike the previous example, this one calls for a polyphonic setting, so turn off mono if it is on.
2. Set two (or more) oscillators to sawtooth and set them to equal balance with the mixer.
3. Use a fine-tuning control to detune the second oscillator slightly to thicken up the sound (to taste!).
4. Bring the cutoff frequency of the LPF to a low setting, perhaps around 1000 Hz (most instruments don't specify the exact cutoff frequency value so experiment until you find something you like). A little resonance is can be added to the filter for additional color.
5. Enable your instrument of choice to allow an envelope generator to control the filter. In the case of the Prophet V you can see that the ENV AMT setting is full.
6. Set the envelope similar to the settings in the previous filter section: fast/moderate attack, slow/moderate decay, 50% sustain level, and slow/moderate release.
7. Set the main amplitude envelope with a fast attack, moderate decay, 50% sustain, and a slow/moderate release.
8. Play with all of these controls to see how they affect the timbre, and store the setting when you come up with something you like!

Subtractive synths are also good for creating percussive elements that may bear only a slight resemblance to a real percussion instrument but can be perfect for rhythm tracks that are not trying to sound authentic. The primary difference between a patch in this category and others we have looked is the fact that percussive sounds do not sustain: Once the attack happens, there is only decay, no sustain or release stage. This calls for a simplified volume envelope that has only attack and decay (in some cases you will find attack and release) or to use a standard ADSR envelope with the sustain level turned all the way down so that only attack and decay are relevant.

Again, any subtractive instrument should be capable of producing a similar sound, but for Sound Design 4.3 we are going to demonstrate creating a synth snare sound using Logic's ES2, a three-VCO instrument with extensive control and sound modification capabilities (of which we will barely scratch the surface).

Sound Design 4.3—Synth Snare

To make this work you need a synth with at least two oscillators, a LPF, and two to three EGs (see Figure 4.6). When emulating a complex sound, it is necessary to design separately the constituent parts that make up the timbre. At its most basic, a snare drum is made up of a resonant body (the drum itself) and metal snares that rattle on the bottom drumhead. In designing this sound we use a pitched sound to emulate the drum, and noise will be adequate for representing the snares. For the drum resonance we use a triangle wave.

Where it gets a little trickier than what we have dealt with so far, the oscillator generating the triangle wave and the one producing the noise will go through a

Figure 4.6
Sound Design 4.3 synth snare using Logic's ES2.

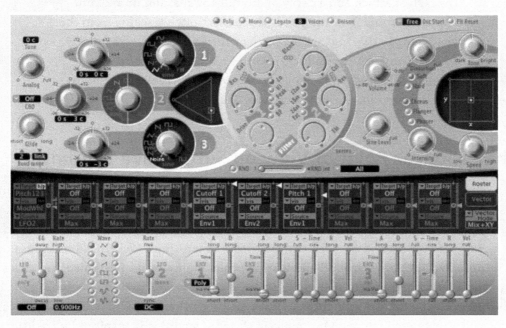

volume envelope to achieve the proper attack and decay we are looking for. A second envelope adds another layer of interest by shaping the *pitch* of the triangle wave (if you listen closely, the pitch of a drum has a tendency to drop as the vibration decays). If available, use a third envelope to shape the cutoff of the LPF so that the brightness attenuates quickly as well.

Steps

1. Set Oscillator #1 to a triangle wave (again, the intention of this sound is to emulate the resonance of the drum itself independent of the snares rattling on the bottom).
2. Set Oscillator #2 to white noise (or just "noise") for the sound of the snares.
3. Use one EG to control the pitch of Oscillator #1 (EG #1 in the preceding example; notice in the matrix how it is routed from Env1 to a Target of Pitch 1). Set the envelope to the fastest attack possible and a decay of around 250–300 ms. Turn sustain all the way down; the placement of release is irrelevant because it will not be active with the sustain off.
4. Use another envelope targeted to the cutoff frequency of the LPF (Env 2, routed to Cut 1+2 in the matrix). Set the envelope's attack to the fastest possible and a very short decay of around 1–2 ms—again, sustain all the way down.
5. Set the LPF cutoff to taste; ours is around 2.5 kHz, the area where a real snare drum has a strong concentration of overtones.
6. Set the overall amplitude envelope (Env 3 in the preceding example) to the fastest attack and a decay of around 150 ms, and no sustain.
7. f your instrument allows you to disconnect pitch tracking from the keyboard, do that and set the pitch of the triangle wave manually to taste, probably around 220 Hz where a standard snare drum resonates. If pitch is tracking, as with our ES2, try different notes on your MIDI controller until you have a fundamental pitch you like.

The Producer Point of View on Subtractive Synthesis

As we learned in the first part of this chapter, subtractive synthesis can produce some pretty interesting sonorities even though it starts from some basic principles and raw material (mainly, geometrically shaped waveforms). The fact that this type of synthesis has existed for several decades and that it has been successfully used in thousands of records and hit songs around the world speaks highly of its flexibility and charming appeal when it comes to electronic arranging and production. As you will learn in this and in all the following chapters, each type of synthesis has its peculiar characteristics and sonic personalities. Our analysis of sonic qualities is inspired by traditional orchestration techniques and sonorities, in which each section, family, and instrument of an orchestra features very different, yet compatible, sonic traits. One of the goals of this book is to make you look at sound design and sound production in a similar way. We often like to compare synthesis to wine. In particular, we like to think of different types of synthesis as different types of wine, in which the grapes, their relative blends, and their basic characteristics can give birth to an almost infinite variety of colors, shapes, and personalities. In this chapter we analyze the sonic qualities of every type of synthesis in relationship to the following standard parameters: overall personality, edginess/smoothness, transparency, thickness, punch, and weight. Looking at sound design through these six categories will enable you to quickly categorize, and sequentially create, any type of sonority.

Subtractive synthesis is a pretty flexible tool that can have several personalities, but it has a few common underlying features. To hear some general audio examples of analog synthesis generated by a subtractive synthesizer, listen to Audio Examples 4.1–4.4. For a quick description of its main sonic characteristics look at Table 4.1.

We recommend having at least one or two subtractive synthesizers as part of your setup to enable you to use some vintage "analog sounds." It is important to have as many sounds and patches available in your studio as possible in order to have a rich palette to choose from. For any electronic-based productions, the vintage sonority offered by

Table 4.1 The Main Sonic Characteristics of Subtractive Patches

Feature	Characteristics
Overall personality	Subtractive synthesis has an overall simple and almost "shy" personality with some definitely naïve features. These traits are probably rooted in the fact that it is one of the oldest and most basic types of synthesis.
Edginess/ smoothness	Subtractive synthesis has a double personality. It can present both edgy and smooth sonorities simultaneously. Pads, some basses, and some leads can have a rich tone with a smooth (almost "gummy" at times) sonic print if constructed from sine and triangular waves, whereas other sounds (such as leads and some basses) can feature aggressive and edgy sonorities if based on sawtooth, square, and white-noise waves.
Transparency	Subtractive synthesis in general has a fairly low level of transparency, mainly because of its frequency-subtraction nature.
Thickness	Most sounds created by subtractive synthesis have a medium-to-high thickness. Its sonic features give an overall sense of depth and intensity. This is particularly true for patches that feature sawtooth and triangular waveforms.
Punch	Subtractive synthesis is capable of delivering sonoroties with a pretty good punch, especially in the mid–low register, making it a good choice for brass and powerful basses that need to support a deep BD.
Weight	Because of its nature, this type of subtractive synthesis can create a variety of patches that feature both heavy and medium weights.

subtractive synthesis is a must-have. Later in this chapter we will listen to and analyze in more detail specific subtractive synthesis patches.

Subtractive Synthesis Hardware and Software Sources and Options

When it comes to subtractive synthesis the options in terms of sources, both hardware and software, are many. Because of its native hardware-based tradition, the first option to consider when "shopping" for subtractive synthesizers is the vintage route. Some of the most important synthesizers that used subtractive synthesis as their main sound-generator techniques are the famous Minimoog, the Oberheim OB series, the Prophet 5 by Sequential Circuits, and the Juno series by Roland. It is still fairly easy to find secondhand models of vintage subtractive synthesizers such as the Moog, Minimoog, and Roland Junos. Even though the vintage hardware solution definitely has its charm, and it can deliver the most authentic sonorities, it also can be problematic in terms of reliability, durability, and convenience. To integrate well with your current DAW and digital equipment, a vintage model must have a built-in MIDI interface. Only original subtractive synthesizers made after 1982 had this option built in, but it usually can be added as a custom-built option. Still, the periodic retuning that some of the vintage machines require can be charming but not very efficient if one is working under a tight deadline. Because of all these issues with the vintage hardware option, we usually recommend considering either new hardware subtractive synthesizers or software synthesizers that emulate the original vintage hardware. See Table 4.2 for a list of the pros and cons of these two options.

In general, we recommend investing in a good selection of subtractive software synthesizers in order to have in your studio a comprehensive sound palette based on these types of synthesis. This will guarantee you the highest level of flexibility without the headache of dealing with vintage machines and without the higher cost associated with

Table 4.2 Subtractive Synthesizers: Comparison of New Hardware and Software Versions

New Hardware Subtractive Synthesizers	**Software Subtractive Synthesizers**
Pros	Pros
• Realistic sound	• Extremely versatile
• Traditional analog warmth	• Extremely expandable and upgradable
• Transportability (for live usage)	• Relatively inexpensive
• Hardware controllers	• One synthesizer can be used in several instances
	• Excellent editing features
	• High fidelity in reproducing vintage sonorities
Cons	Cons
• Bulky	• They require a computer with fast CPU
• Advanced editing can be difficult and sometimes cumbersome	• Sometimes they lack the warmth of their hardware counterpart
• Limited upgradability	
• Expensive	

newer hardware synthesizers. Eventually you can look at expanding your palette with one or two hardware options if budget and conditions allow.

Which Subtractive Device or Software Synthesizer is Right for My Productions?

As you can see, when it comes to choosing the right gear or sound source for your productions the options are many, and determining the best choice can seem overwhelming. For an overview of the options available and their main features, see Tables 4.3, 4.4, and 4.5, where we present a selection of our favorite subtractive synthesizers (both hardware and software, vintage and new).

Table 4.3 Selected Vintage Hardware Subtractive Synthesizers

Vintage Hardware Subtractive Synthesizers	**Features**
Arp 2600 Arp Odyssey MK III	The Arp 2600 is a semimodular monophonic synthesizer that has been used for more than 20 years in a variety of styles. Its strong feature is its extreme flexibility and overall sound quality. The Arp Odyssey is a compact version of the 2600, offering a scaled-down version of the 2600 engine in a more portable package.
Moog Prodigy Minimoog	The Prodigy is a monophonic synthesizer particularly useful for analog bass patches. It is considered the "cheap" version of the Minimoog, which was also a monophonic synthesizer but that featured the famous filter designed by Robert Moog. Its signature sounds are mainly leads and bass patches that have been used in a variety of styles by such artists as Blondie, Keith Emerson, Rick Wakeman, Toto, and Joe Zawinul.
Prophet 5	The Prophet 5 by legendary Sequential Circuits was the symbol of analog synthesis in the early 1980s. It has no MIDI (the later model Prophet 600 was equipped with a MIDI interface), and it suffers from the "bugs" of early analog synthesizers (mainly unstable tuning). Nevertheless, its flexibility and signature pads, strings, and basses made it one of the top analog synthesizers of the 1980s.
Roland 106 Roland Juno 60	The Juno 106 is still considered "the" analog synthesizer for dance and contemporary production. A polyphonic synthesizer with MIDI ports (sixteen MIDI channels), the Juno 106 is easy to program and provides a great variety of sonorities, ranging from rich pads to edgy basses. Such artists as Vangelis, BT, Moby, and Depeche Mode made this their signature synthesizer. The Juno 60 is the early version of the 106, with basically the same engine but without MIDI I/O.

(continued)

Table 4.3 Continued

Vintage Hardware Subtractive Synthesizers	Features
Korg PolySix	In the same category of the Juno series this Korg polyphonic synthesizer was among a wave of fairly affordable machines to come out in the early 1980s. Its most famous patches are brassy and chorus pads. It was made famous by such artists as Kitaro, Tears for Fears, and Keith Emerson.
Yamaha CS-15	The CS-15 is a monophonic synthesizer that was very strongly built. It is very flexible, and it can create a variety of sounds. We like it especially for its leads, bass and sound effects patches.
Siel Cruise Siel Opera 6	Siel was an Italian maker that created a few analog synthesizers in the late 1970s and early 1980s. The Cruise, which has no MIDI ports, was one of our favorites at the time because of its easy programming, good live performance controls, and smooth sonorities. Particularly interesting were the woodwind sounds and orchestral sounds in general. Its "big brother" was the Opera 6 that had a unique sonority and was particularly appreciated for its pad and string sounds. Both are pretty hard to find, but worth the search!

Table 4.4 Selected New Hardware Subtractive Synthesizers

New Hardware Subtractive Synthesizers	Features
Roland JD-X-A	The JD-X-A is a hybrid analog–digital synthesizer. It takes advantage of all of the innovations that have happened in synthesis in the last 20 years while retaining the "analog" character and design of the old machines.
Moog Voyager Moog Little Phatty	These two monophonic synthesizers are a great source of vintage sonorities with a contemporary twist. The Voyager is great for bass and leads; the Little Phatty provides extensive editing and arpeggio options.
Nord Lead A1	The Nord Lead series provides a lot of flexibility with a modern sonority strongly linked to the tradition of analog subtractive synthesis. This model is polyphonic, carrying the tradition of vintage synthesizers like the Prophet 5 into the twenty-first century.

Table 4.5 Selected Software Subtractive Synthesizers

Software Subtractive Synthesizers	Features
Arturia V Collection Classic	Arturia has been in the business of vintage synthesizer emulation for many years. It offers among the best classic emulations available with extremely realistic sonic reproduction. This collection includes the Jupiter 8V, Prophet V, ARP2600 V, and the Minimoog V.
Applied Acoustics Systems–Tassman	Tassman is a completely modular software synthesizer capable of reproducing practically all the most important types of synthesis available through an easy interface and editors. You can build your own subtractive synthesizers (and many more) by simply adding the modules that are specific to this type of synthesis. It is very flexible and fun to work with.
Logic Pro X – ES1, ES2, Retro Synth	These three software synthesizers are bundled with Logic Pro. They offer an excellent resource for creating both vintage and contemporary subtractive patches. Their signature sounds are leads, basses, and pads.
Korg Legacy Analog Collection	This collection of vintage Korg analog synthesizers features three vintage gems: MS-20, Polysix, and MonoPoly.

It is important to keep in mind that, even for types of synthesis that have been utilized in production for decades, as is the case for subtractive synthesis, new hardware and especially new software synthesizers are being created regularly. Therefore we highly recommend that you constantly update your sound sources and stay up-to-date with the latest software synthesizers in order to always keep your sound palette fresh and original.

Sonic Categories and Subtractive Patches

One of the most important aspects to consider when discussing any type of synthesis and sound generation is the sound quality and the overall sonority they are capable of producing. Starting in this chapter, and continuing in those that follow, we analyze in detail the specific patches and sound categories that each type of synthesis is best suited to produce, and which sonorities constitute their sonic signature. In general, subtractive synthesis is particularly suited for creating low and deep analog basses, analog leads, rich and thick pads, and to a certain extent some acoustic sounds like synth strings, brass, and woodwinds. These are among the sounds and patches that made subtractive synthesis famous and that are still largely used in contemporary productions.

Subtractive synthesis, because of its limited basic waveforms, is usually not suited to faithfully recreating acoustic instruments, even though it can be somewhat effective in producing synthesized strings and woodwinds. Although this type of synthesizer was used in the 1970s and 1980s to sequence some orchestral parts, today it can't compete with more advanced sample-based synthesis options.

Bass

Subtractive bass patches are probably among the most important sounds that have characterized this type of synthesis. The legendary Moog bass sounds marked an entire era of contemporary production in the late 1970s and early 1980s. Subtractive basses can be grouped into two categories: sonorities that are smooth and deep, and sounds that are edgy and punchy. The main qualities of these two categories depend primarily on the shape of the waveform from which they originated. Generally speaking, a sine or triangular wave will give you smoother and deeper sounding basses, whereas a sawtooth or square wave will translate into edgier and punchier basses. As you learned in the previous chapters, the envelope of the amplifier has a big impact on the sonic quality of a sound. This is particularly true for subtractive patches in which you want to achieve different shades of smoothness/edginess, as is the case for a bass patch. For most contemporary styles the key element of a bass part resides in "locking in" with the bass drum (BD) part; therefore, its sonic characteristics in terms of "punchiness" and smoothness play a key role. See Table 4.6 for an in-depth look at how to shape subtractive bass patches with different sonic characters.

Listen to Audio Examples 4.5–4.8 to hear the aforementioned examples of analog bass patches in action.

As the main role of the synth bass is to complement and interlock with the Bass Drum (BD), try to match a deep bass drum with a slightly lighter and edgier bass patch. Then, vice versa, combine a deeper bass patch with an edgier BD sound to achieve a satisfactory balance.

Sequencing and Production Techniques for Analog Bass

The synthesized bass in general, and the analog subtractive bass in particular, usually is linked to electronic styles such as disco, rhythm and blues (R&B), electronic soul, house, trance, and rave. Its use is deeply rooted in late 1970s and early 1980s European dance music production. Its primary role is to support the BD (usually accentuating each quarter note) while, at the same time, adding some rhythmic patterns that have the dual function of outlining the harmony and creating rhythmic drive and variations. The patterns are usually repetitive, with 8- or 16-bar cycles. It is often used with an arpeggiator in order to give the impression of mechanical repetition and drive (Figure 4.7).

It is not uncommon to have the notes of a subtractive bass part set to the same velocity or limited to two or three velocity values in order to re-create a "bass machine" such as the vintage Roland TB-303. You can, in fact, use the pencil entry tool of your DAW to sequence the bass part or, alternatively, use a step sequencer if your DAW provides one (Figure 4.8).

To add a bit more of a human touch to your bass parts you can use a combination of Pitch Bend and Modulation wheel. Pitch Bend allows you to insert some nice variations when "entering" or "exiting" a sustained note. We particularly like to use it at the end of a long sustained bass note to mitigate the short release that most bass patches have. Use it sparsely and with taste—too much of it can easily spoil your production. As noted in Table 4.6, we strongly advise always constructing subtractive bass patches in monophonic mode. This technique has a few advantages. The primary advantage is that

Table 4.6 Key Elements of Subtractive Bass Patches

Type of Bass Patch	Key Elements	Tips
Smooth deep bass	Sine waveform Short attack time LP 12 filter @ ~1000 Hz Uses monophonic settings	The sine wave provides a nice, smooth contour that is further mitigated by the LPF. Play with the filter to add more or less edginess to the patch. Also, try to use a mix of sine and triangular waves to give a slightly edgier and more open character to the patch. Use a monophonic setting to add a realistic bass feel.
Smooth light bass	Mix of sine and triangular waveforms Short attack time LP 12 filter @ ~2000 Hz HP 12 filter @ ~80 Hz Uses monophonic settings	The addition of a triangular wave to the sine adds a bit more definition, and the HPF gives a lighter sonic character.
Edgy deep bass	Sawtooth waveform Short attack time LP 12 filter @ ~800 Hz Uses monophonic settings	The sawtooth waveform will give you a very edgy and sharp sound. Combined with the LPF, it will give you a deep and punchy bass that will complement well any electronic bass drum.
Edgy light bass	Either square or sawtooth waveforms (or any combination of the two) Short attack time LP 12 filter @ ~1000 Hz Uses monophonic settings	The combination of the square waveform (which gives a lighter sonic feel in general) with the deeper sawtooth creates an effective combination for a slightly less deep bass patch.

Figure 4.7
Typical synth bass patterns for the subtractive bass.

Creating Sounds from Scratch

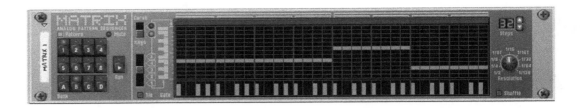

Figure 4.8 Step sequencer in Reason.

you will be able to create some nice legato lines that will add some nice subtle nuances to the bass part. The fact that the synthesizer works in monophonic mode implies that if a note is kept pressed while the next one is played the second note won't trigger the attack portion of the amplifier envelope, resulting in a smoother and less edgy sound. Another advantage is that you will avoid playing double-stops (two notes at the same time), which are not very common in bass parts for the aforementioned music styles.

Mixing the Analog Bass

The mix stage of any synthesized sound is crucial to achieving the highest possible production quality. As is the case with acoustic instruments, each electronic sound has its own specific techniques and rules. Analog subtractive bass patches follow, for the most part, the basic rule of mixing that we apply to any bass track. Remember that the main goal for bass tracks is to have a deep but defined sound that works perfectly in synergy with the BD to provide a punchy, solid, and unified bass frequency experience. Through the use of the filter and additional equalizers, try to shift the frequency window of the bass patch so that it fits and complements the BD. With the BD and bass part soloed, adjust the equalization of both so that they complement each other. For example, if the BD is tight and punchy, the bass can extend a bit lower and fill the frequency space below the BD. Conversely, if the BD is deep and low you can have the bass covering the low–mid frequency in order to add more punch. For equalization settings for the analog subtractive bass see Table 4.7.

Pan the bass in the middle and avoid extreme panning settings. Also, to re-create the original analog sonority you can keep the patch in mono. The subtractive bass usually doesn't need a lot of effects at the mixing stage. You definitely don't want to add any reverb, as this would reduce the "punchiness" and definition that are typical of these

Table 4.7 Key Equalization Frequencies That Apply to Analog Subtractive Bass

Frequencies	Application	Comments
60–80 Hz	Boost to add fullness	This range gives you a fuller sonority without over accenting muddiness.
200–300 Hz	Cut to reduce muddiness	Use with discretion. By cutting too much you might reduce the fullness of the bass track; by boosting too much you might increase its muddiness.
400–600 Hz	Boost to add presence and clarity to bass	
1.4–1.5 kHz	Boost for intelligibility	

Table 4.8 Listening Corner: Subtractive Bass

Artist(s)/Track/Genre	Featured Bass Lines
Eurhythmics, "Here Comes the Rain Again," from "Touch," 1983, RCA Records–Synthpop	The track features a deep and punchy analog bass.
The Promise, "When in Rome," 1987, Virgin–Synthpop	A mellow and deep bass with a typical rhythmic part.
McBrian, "Loving You Forever," 1989, High Energy–Italo (Eurodisco)	A perfect example of punchy subtractive bass with a typical "8th–16th–16" note pattern.
Lipps, Inc., "Funkytown," 1979, Casablanca Records–Disco	A simple analog bass with a simple 8th note pattern.
Harold Faltermeyer, "Axel F," "Beverly Hills Cop Soundtrack," 1985, MCA Records–Film	This classic bass line features a smooth subtractive bass that locks in perfectly with the bass drum.

sonorities. Sometimes a small amount of chorus can enhance the thickness of the bass, especially for parts that are not very rhythmic and in which notes are mostly sustained rather than short and staccato. For some examples of productions featuring the subtractive bass visit the Listening Corner table (Table 4.8).

Pads

Subtractive pad patches can vary considerably in weight, color, texture, edginess, and overtime movement. Some of our favorite pads are programmed on subtractive synthesizers because of their warmth, thickness, and high playability. To understand how a pad patch can be constructed or edited to your needs, it is important to understand the role of a pad in contemporary production. Generally, pads can have two main functions: a primary function (we refer to these pads as "featured") in which the pad is more upfront in the mix (it is placed in the foreground and alternated between a lead role and a harmonic role), and a secondary function in which the pad plays a background role and is usually mixed in the background. Featured pads usually are richer in tone, brighter, and also use LFOs, modulation envelopes, and filter envelopes to create a sense of constant motion (i.e., a sweeping-filter effect). These types of pads have been brought to the attention of the public by such artists as Vangelis and Kitaro, in whose music the pads themselves constitute the main thread of a piece. Background pads, on the other hand, are generally more subdued, darker, and mellower, and serve a supporting function, usually to the harmony of the piece. Of course, there are several variations between these two main categories, but you will find yourself using these two extremes as your reference points when programming pads for your contemporary production. See Table 4.9 for a detailed list of parameters that identify the most common types of pads in subtractive synthesis.

Listen to Audio Examples 4.9–4.13 to hear the examples of analog pad patches listed in Table 4.9.

Table 4.9 Key Elements of Subtractive Pad Patches

Type of Pad Patch	Key Elements	Tips
Rich and thick pad	2 OSCs OSC 1 pulse width OSC 2 sawtooth or pulse slightly detuned from OSC 1 Medium attack time Full sustain Medium to long release LP 12 filter @ ~2000 Hz Uses polyphonic settings Adds some resonance to the filter of OSC 1 to make the pad richer	This type of pad usually plays a featured/prominent role in the piece. For added thickness, set OSC 2 an octave lower than OSC 1. To add even more richness to the patch try to activate the envelope of the filter with a very slow attack, medium decay, full sustain, and medium release.
Rich and light pad	2 OSCs OSC 1 PW OSC 2 SAW slightly detuned from OSC 1 Medium attack time Full sustain Medium to long release LP 12 filter between 7000 and 5000 Hz Uses polyphonic settings Adds some resonance to the filter of OSC 1 to make the pad richer	This type of pad usually plays a featured/prominent role in the piece. The higher frequency of the LPF gives a brighter and lighter color to the pads.
Rich and edgy pad	2 OSCs OSC 1 SAW OSC 2 SAW or pulse slightly detuned from OSC 1 Very short attack time Full sustain Medium to long release LP 12 filter between 7000 and 5000 Hz Uses polyphonic settings Adds some resonance to the filter of OSC 1 to make the pad richer	This type of pad usually plays a featured/prominent role in the piece. For added thickness set OSC 2 an octave lower than OSC 1. To add even more richness to the patch try to activate the envelope of the filter with a very slow attack, medium decay, full sustain, and medium release. These settings can easily bridge between a pad and a subtractive brass patch. The brighter the filter section and the shorter the attack, the closer the patch will be to an analog brass sonority.

Smooth and airy pad	2 OSCs OSC 1 sine OSC 2 triangular slightly detuned from OSC 1 Medium to long attack time Full sustain Medium to long release LP 12 filter between 9000–8000Hz Use polyphonic settings Add some resonance to the filter of OSC 1 Filter envelope: medium attack, medium–short release	This type of pad usually plays a background/secondary role in the piece. For added thickness set OSC 2 an octave lower than OSC 1. To add even more richness to the patch try to activate the envelope of the filter with a medium attack, medium decay, full sustain, and short release. Add reverb, delay, and a touch of chorus for extra smoothness.
Smooth and opaque pad	2 OSCs OSC 1 sine OSC 2 triangular slightly detuned from OSC 1 Medium to long attack time Full sustain Medium to long release LP 12 filter between 9000 and 8000 Hz Uses polyphonic settings Adds some resonance to the filter of OSC 1 Filter envelope: medium attack, medium–short release	This type of pad usually plays a background/secondary role in the piece.

As noted at the beginning of this section, pads can vary in color and shape. This is one of the areas in which we invite you to experiment and try to be adventurous. Always think of the purpose of the pad in the arrangement first. Is it going to serve a featured/prominent role or more of a background role? With its purpose in mind you can shape the pad as you wish with the sonic characteristics that will best serve its purpose.

Sequencing and Production Techniques for Analog Pads

Sequencing analog subtractive pads is pretty straightforward. Usually a good keyboard controller will do the trick. As an alternative we sometimes like to use a GMC in order to use original voicings and different dynamics. This type of controller works particularly well for featured pads, in which the ability to use different dynamics for each note of the voicing allows you to create more variations and interesting lines. Typical subtractive pad parts feature either a simple triadic voicing to outline the harmonic progression, or a one- or two-note voicing (usually in either octaves or fifths) in the low register to create the typical low pad effect.

When it comes to creating your own pads, one technique that we highly recommend is layering. This technique enables you to create complex and original pads in a matter of minutes. Although we particularly like to layer pads and sounds from different types of synthesis, this approach is very effective also with any type of subtractive pad. In fact, because of the simplicity of the basic waveforms available in this type of synthesis, layered pads are extremely rewarding when applied to subtractive synthesis. Depending on your DAW, the technique of layering different sounds from different sound sources will vary, but the concept is the same: Combine two or more pads so that they are triggered by the same MIDI track simultaneously. Assign the output of one of your MIDI tracks to two or more software synthesizers, and load a different patch on each software synthesizer. This will give you complete control over the volume and balance of each source, but you will have to record and edit only the MIDI data on one track. The key aspect of this technique, of course, is the choice of the pads that you will layer together. We usually like to combine complementary pads that complete each other without stepping on each other. For example, try layering a deep and dark pad patch with an edgy and shiny one. The result will be a completely new sonority that will cover a wider area of the ensemble. Another example would be the layering of a mellow and slow pad with a more decisive and punchy one. The result will be a fuller pad with a nice attack. As you can see, here the sky is the limit.

Mixing Subtractive Pads

Mixing pads gives you a good amount of freedom in terms of both equalization and use of effects. Once again, different pad functions call for different mixing approaches. Let's analyze first the key features of a mix for background pads. For this category you can use the equalizer and the filter to narrow down a frequency range that fits nicely with the other instruments of the arrangement. Usually you will need a background pad to play nicely with another harmonic instrument (piano, organ, or guitar) and with the vocals. The key point here is to carve a nice space in the frequency spectrum for the pad to fill without overstepping other instruments. Avoid boosting the bass frequencies excessively, as this could clutter the low end and interfere with the bass and BD. Use a HPF to cut the very low frequencies. Also, avoid making it too bright, because a high content of high frequencies will take away the subtractive warmth and conflict with the background function of the pad. Remember, this pad takes a "supporting role," not a featured one. In terms of effects, a nice reverb with a length between 1.7 and 2 s usually is a good starting point. Try not to overuse the reverb, as this will take away definition. On the other hand, if you want to re-create a retro mix with reference to the 1980s and 1990s, using a longer reverb will help you. To thicken the pad a bit you can use modulation effects like chorus or ensemble. These effects are absolutely "legitimate" for analog background pads, as they were often available as add-on effects on the actual hardware machines. In terms of panning, pads generally seat better in a mix if they are spread over the stereo image. This will avoid clutter in the center of the mix and also will allow you to create interesting panning effects with pan automation. To create a sound that is closer to the original subtractive vintage machines you can keep the pad in mono, but we recommend this technique only if absolutely necessary for stylistic reasons.

If you are mixing a featured pad you have more creative freedom. The equalization settings usually tend to be more permissive and original. Make sure that the pad has a nice low end with a solid frequency range between 80 and 180 Hz. The mid–high and high ends should be clear, defined, and crisp (but not distorted). Reverb settings can vary on the part (usually less rhythmically active parts can have longer reverb times), but a length of 1.9 to 2.3 s is a good starting point. We usually like to use modulation effects like flanger, ensemble, or chorus with slow cycles to enhance the variation of the patch over time. In terms of panning, try to open the pad in a nice wide stereo image and add a subtle pan automation to bring the part to life. For some examples of productions featuring subtractive pads, visit the Listening Corner table (Table 4.10).

Table 4.10 Listening Corner: Subtractive Pads

Artist(s)/Track/Genre	Featured Pads
Vangelis, "Chariots of Fire," 1981, Polydor–New Wave/Soundtrack	This track features a deep steady pad that evolves around 1'40" into a more active and featured pad.
Jean Michel Jarre, "Oxygene Part IV," 1977, Disques Dreyfus/Polydor–New Wave	A breathy and evolving pad with the envelope of the filter programmed to open, a slow attack bass, and a typical rhythmic part.
Kitaro, "Flight," from "World of Kitaro," 1981, Sound Design–New Age	A deep and rich brassy pad that starts with a featured function, then assumes a more background role when the other instruments join in.
Giorgio Moroder, "The Autopsy," from "Cat People Soundtrack," MCA Records, Backstreet Records–Film	A deep and dark pad opens the track. Its sonority is bleak and rich with a constantly changing color.
The Goblin, "Patrick M32," from "Patrick Soundtrack," 1979–Film	A light and airy pad opens the track. At 16" the pad gets richer and slightly thicker.
Sylvian/Fripp, "Bringing Down the Light," from "The First Day," 1993, Virgin–Ambient	A deep pad functions as bass, while a light and dreamy pad is layered on top, serving almost as lead.

Creating Sounds from Scratch

Strings and Brass

Directly related to the generic category of pads are two other typical subtractive patches: strings and brass. Although they both present some sonic characteristics that differ from those of the typical pad, we include them in this section because their arranging and production functions are very often comparable to those of the typical pad.

Analog subtractive strings were among the sonic signatures of the early analog synthesizers. Polyphonic synthesizers like the Prophet 5, the ARP String Ensemble, and the Oberheim OB-Xa made the subtractive string sonority a must-have for any producer in the early 1980s.

A basic string patch can be constructed starting from two oscillators set to pulse wave, with a LFO set to control the phase of one of the two OSCs in order to create a richer sound that gives the impression of a larger ensemble. Slightly detuning the two OSCs symmetrically (e.g., ± 8 cents) also will add a nice ensemble effect. Attack should be set to medium or slow, decay and sustain should be set to full, and release should be set to medium. To mellow out the edginess of the pulse waves use a twenty-four-pole LPF with the cutoff frequency set between 9000 and 8000 Hz. To add a deeper effect, detune OSC 2 an octave lower (Figure 4.9).

For an example of a subtractive strings patch, listen to Audio Example 4.14.

Adding a nice reverb also will help to create a more natural environment and add some realism to the patch. For strings you can use a reverb length of 2.3 to 2.8 s. It is important to remember that subtractive strings won't compete with any sampled or physically modeled patches that you can create with modern software synthesizers. Instead, they can be very useful in recreating a warm retro sound for any electronic production.

Analog subtractive brass patches are probably among the most well-known sounds since the early 1980s. Literally hundreds of dance, pop-rock, and electronic tracks have featured subtractive brass patches. Consider the familiar intro section of "Jump" by the rock group Van Halen, in which the synthesizer played a polyphonic triadic part over a C pedal. That patch was, in fact, programmed and played on an Oberheim OB-Xa. The main characteristic of a subtractive brass patch is its punchiness and richness. The term "brass" covers a very wide sonic landscape, ranging from mellow and smooth sonorities

Figure 4.9
Example of a subtractive string patch setting in Reason's Subtractor.

Figure 4.10
Example of a typical subtractive brass part.

all the way up to bright and edgy sounds. We find that, overall, the most appropriate use of subtractive brass patches is for contemporary edgy and bright electronic productions (Audio Example 4.15). These patches fit well in any dance style for which a punchy and rich sound is needed. A typical punchy synth brass part would look like the one shown in Figure 4.10.

These sonorities feature two oscillators with sawtooth waveforms with a LPF set around 3000 Hz (if you need punchier and brighter brass, raise the cutoff frequency of the filter), fairly short attack, around 75% decays and sustain, and around 50% release (Figure 4.11).

Although for more acoustic, orchestral, and classical styles we will use sample-based and physical modeling synthesis techniques later in the book, you can get some fairly convincing orchestral brass by using two oscillators with sawtooth waveforms and a LPF with a lower cutoff frequency of around 5000 Hz and a slightly slower attack.

A reverb with a length between 2.0 and 2.3 s provides a nice ambience for analog subtractive brass patches, and a little bit of chorus can enhance the ensemble features of these patches. For some examples of productions featuring subtractive strings and brass visit the Listening Corner table (Table 4.11).

Leads

The third type of patch, and the one that has made subtractive synthesis so popular in contemporary production, is "leads." Subtractive leads seem to have timeless appeal, as literally every contemporary dance, house, trance, and electronic production since the

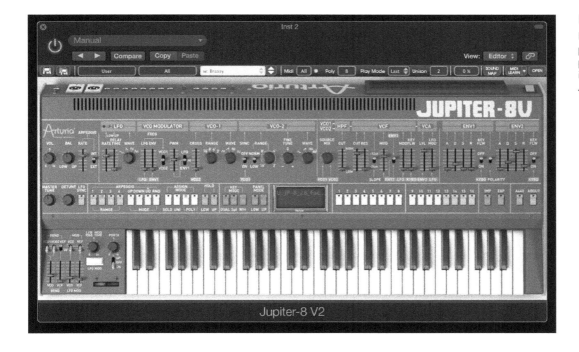

Figure 4.11
Example of a punchy brass patch setting in Arturia's Jupiter 8V.

Table 4.11 Listening Corner: Subtractive Strings and Brass

Artist(s)/Track/Genre	Comments
Kitaro, "Flight," from "World of Kitaro," 1981, Sound Design–New Age	A rich and deep brass solo line opens the track. The sound is punchy, yet reminiscent of a classical French horn section.
Animotion, "Obsession," 1983, Mercury Records, Polydor Records–Synthpop	At 2'20" a punchy polyphonic brass patch underlines the harmony and at the same time creates rhythmic syncopation.
Frankie Goes to Hollywood, "Relax," 1983, ZTT–Synthpop	A punchy and edgy brass patch makes its entrance at 45" with a classic rhythmic riff.
Van Halen, "Jump," from "1984," 1984, Warner Bros. –Rock	The rich brass patch has become a classic for subtractive synthesis. The synth is an Oberheim OB-Xa.
Elton John, "Someone Saved My Life Tonight" from "Captain Fantastic and the Brown Dirt Cowboy," 1975, MCA–Pop	A high string patch played by an ARP String Ensemble is featured at around 5'20."
The Buggles, "Video Killed the Radio Star," 1979, Island–Pop	A descending line in the high register is played by the ARP Strings Ensemble during the famous chorus.

mid-1980s has featured one or more synth leads. As we learned for basses and pads, leads can vary quite dramatically in color and edginess. In addition, the use of such effects as saturation, distortion, modulation, reverb, and delay can further increase their sonic variety. Leads can be divided into two main types: mellow and edgy. The former is usually associated with R&B, some hip-hop, and even electronic pop, whereas the latter is characteristic of such genres as house, techno, and trance. Leads are always monophonic patches that can take great advantage of *portamento*, modulation, and pitch bend in order to enhance control of the performance. In Table 4.12 we review the key elements of subtractive leads.

Sequencing and Production Techniques for Subtractive Leads

Sequencing a lead part can be done effectively through a standard MIDI keyboard controller. Usually we use a synth-key action to sequence lead parts because it enables us to play with a higher agility level and because it re-creates the original synth-key action feel of the vintage synthesizers. To improve playability and expressivity we highly recommend employing a Breath Controller, as we saw in Chapter 3. In general, try not to quantize your lead parts unless absolutely necessary. The warmth of the subtractive synthesis leads really comes out when the lead part naturally flows without the constriction of strict grid-based quantization. Listen to Audio Examples 4.16 and 4.17 to hear the two main types of leads.

Table 4.12 Key Elements of Subtractive Lead Patches

Type of Lead Patch	Key Elements	Tips
Mellow and smooth lead	1 OSC OSC 1 saw or triangular wave Short to medium attack time Full decay and sustain Short to medium–short release LP 24-dB filter with a low cutoff frequency Uses monophonic settings Adds very little resonance to the filter of OSC 1 Sets the modulation envelope to control the cutoff frequency of the filter with a short attack Short *portamento* (glide)	This type of lead is smooth and playful. The lower the frequency of the filter, the smoother and more mellow the patch will sound. Add a delay and a medium reverb (1.8–2.2 s reverb length) to make the patch more melodic.
Sharp and edgy lead	2 OSCs OSC 1 pulse wave OSC 2 saw or pulse Short attack time Full decay and sustain Short to medium–short release LP 12-dB filter between 5000 and 4000 Hz Uses monophonic settings Add some resonance to the filter of OSC 1 to make the lead richer and more resonant Set the modulation envelope to control the cutoff frequency of the filter with a decay of ~1800 ms Short *portamento* (glide)	This type of lead is edgy and punchy. The higher the frequency of the filter, the brighter and more sparkly the patch will sound. Add a delay and a short reverb (1.6–1.9 s reverb length) to create the impression of a longer release. You can also try to use a gated reverb for a more retro sound. Add a saturation or a distortion to add more "byte." For a grungier sound add a bit of noise oscillator.

Mixing the Subtractive Leads

Subtractive leads can be panned freely, either in the center or slightly on either side of the stereo image. In terms of equalization, we recommend using the built-in filter of your synthesizer instead of an external plug-in. This will guarantee you a more realistic and truthful vintage sound. Make sure to keep a nice warm sound, especially if you are using a digital plug-in to synthesize your lead sounds. As we mentioned earlier, leads can benefit from using such time-based effects as delay and reverb. For both effects use either some vintage hardware gear or, preferably, plug-ins that can emulate such analog effects as tape delay and spring reverb. This will improve the legitimacy and vintage sound of your productions. For some examples of productions featuring subtractive leads visit Table 4.13.

Table 4.13 Listening Corner: Subtractive Leads

Artist(s)/Track/Genre	Featured Leads
Dj Cor Fijneman, "Venus (Tiesto Mix)," 2008, Black Hole recordings–Techno	This track features a high-power lead with a rich and powerful timbre.
Vangelis, "Hymn," from "Themes," 1983, Polydor–New Age	A very famous theme in which a light and airy lead with a sharp attack plays a dreamy melody.
Tangerine Dream, "Exit," from "Exit," 1981, Virgin–New Wave	A sharp lead with a long release that combines two complementary sonorities.

Exercises

4.1 Program two different subtractive bass patches, one deep and edgy, the other with a smoother sonority.

4.2 Using the two patches created in Exercise 4.1, sequence sixteen bars of two different bass patterns, taking as a model those illustrated earlier in this chapter.

4.3 Take a bass drum of your choice, sequence a simple four-on-the-floor part under the bass parts you create in Exercise 4.2, and fine-tune the bass patches to create a solid punchy synergy between bass drum and bass.

4.4 Program a featured pad with a rich and over-time changing sonority.

4.5 Program a background pad with a light and airy sonority.

4.6 Combine the bass drum, one of the bass lines, and the pad created in the previous two exercises into a cohesive 32-bar sequence.

4.7 Program a lead patch that fits with the sequence created in Exercise 4.6 and compose a simple repetitive riff.

4.8 Mix the short piece you just created, adding appropriate reverb on the pads and delay on the lead.

5

Frequency Modulation Synthesis

Introduction

Frequency modulation (FM) is elegant in its simplicity, yet impressive in the complexity that lies within. It is less intuitive for sound design and therefore intimidating to many who may have unexplored FM instruments in their collection. The Yamaha DX7 (see Figure 5.1)—the runaway success of the 1980s that introduced FM synthesis to the masses—came with a great collection of presets, but few players dug into the editing capabilities that formed those patches because it was *so* different than the subtractive synthesis they were familiar with. Our goal here is to help you get over the initial learning curve so that you will have confidence to explore and design great FM sounds of your own.

How FM Works

For starters, FM synthesis works in the same manner as FM radio transmissions. The frequency you select on the radio dial is called the *carrier* frequency, and it is being modulated by the appropriately named *modulator* signal, which is the audio program being broadcast. Conventionally in FM synthesis, the carrier is a sine wave and the modulator is one or several sine waves (see Figure 5.2).

At its most basic, FM synthesis requires only two sine waves. But rather than adding them together, as would be the result when two VCOs are summed in a mixer, the carrier is being frequency modulated in a fashion similar to how the LFO is used to make vibrato. However, unlike the LFO that generates *infrasonic* waves (those below the audible frequency range), the modulator for FM is coming from a VCO producing sine waves

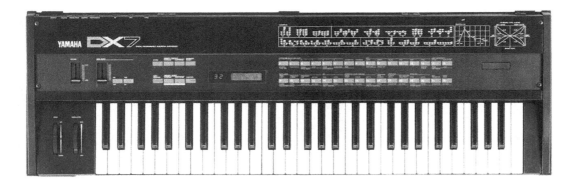

Figure 5.1
The Yamaha DX7 (photo courtesy of Yamaha Corporation of America).

Figure 5.2

AM vs. FM. Left: a sine-wave carrier has been amplitude modulated by a sine-wave modulator; right: a sine-wave carrier has been frequency modulated by a sine-wave modulator (x axis, time; y axis, amplitude/intensity).

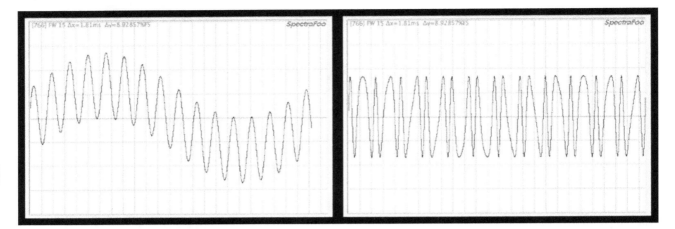

within the audible spectrum. A curious thing happens when the modulator shifts from the infrasonic territory into the audible range: Rather than continuing to hear a vibrato effect, instead we hear added frequencies called *sidebands* that mix with the carrier. With the modulator set to a low level, the result will comprise the carrier frequency plus the modulator's frequency for one sideband and the carrier frequency minus the modulator's frequency for the other sideband—this is referred to as the *sum* and *difference*.

As the intensity of the modulation signal is increased, more sidebands are generated at the sum and difference of the carrier *and* a potentially infinite number of integer multiples (2x, 3x, 4x, etc.) of the modulator. If a negative number results from a "difference" calculation it will still appear in the spectrum with its phase inverted at the equivalent positive value (yikes!).

Figure 5.3 shows a sine wave carrier at 500 Hz being modulated by a sine-wave modulator at 250 Hz. With a relatively low modulator level, only the sum and difference values of 750 Hz (500 + 250) and 250 Hz (500 − 250) are present. As the level of the modulator increases, more sidebands appear. Table 5.1 shows the predictable pattern of sidebands as the modulation level is increased.

Figure 5.3

Spectral analysis comparison. Left: the carrier of pitch C5 (522 Hz) with a modulator introduced at a low level set to C4 (261 Hz) producing sum and difference sidebands. Right: the same carrier modulated at a higher level resulting in multiple sidebands across a wider frequency range.

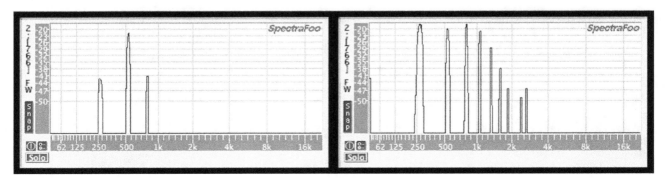

Table 5.1 Calculating FM Sidebands

Carrier Frequency (C)	Modulator Frequency (M)
C + M	C − M
C + 2M	C − 2M
C + 3M	C − 3M
C + 4M	C − 4M*

*etc., as modulation intensity increases.

Does your hair hurt yet? The good news here is that you don't need to go through the math to design sounds with FM, but the basics help you gain an understanding of the underlying principles at play.

To generalize a bit: First, the more of the modulator added, the more complex the resulting spectrum becomes. Second, when there is an integer–ratio relationship between the carrier and the modulator, the resulting timbre will be more harmonic in nature; conversely, noninteger ratios result in timbres that contain more inharmonic frequencies that produce noise. The harmonic timbres in FM tend to be bell-like and have a purity that was not available in the subtractive instruments that preceded, thus the enormous popularity of FM when this new, vast palette of sounds became available. The FM electric piano with its tine-like quality became all the rage in the 1980s and was used extensively (overused?) on pop recordings of the era.

Given the mathematical predictability of FM it is understandable that it was not utilized much prior to the advent of digital oscillators. FM was *possible* with analog instruments—and it's very likely that, either intentionally or by accident, many patch connections were made on modular synths from the output of one VCO to the modulation input of another, with the resulting sound considered either an "oops" or something interesting but nonsensical. Analog oscillators were inherently unstable; their pitch was imprecise to begin with and often drifted as the instrument heated up. As a result, recreating a FM patch or having it remain stable while performing was nearly impossible. Digital oscillators provided the pitch stability necessary to create sounds with predictable and repeatable results.

Algorithms

The timbral complexity offered by a single carrier and modulator is remarkable but merely the starting point for FM. The range of sounds offered by the DX7, for example, and instruments that would follow benefit from chaining a sequence of modifiers in series and/or parallel to two or more carriers so that the same kind of layering that we used in subtractive synthesis would be possible here as well. For the DX7, Yamaha incorporated 32 *algorithms* that contain the essential carrier–modulator combinations for producing a wide range of sounds (see Figure 5.4)

Figure 5.4

Close-up photograph of the DX7 top panel's left-hand side detailing its Algorithms 1–9 and 19–25.

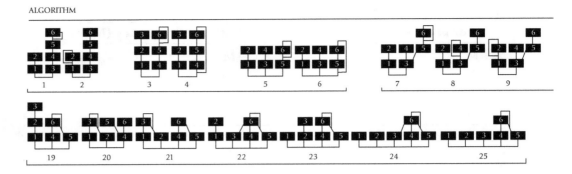

Figure 5.5

Yamaha DX7 Algorithm 1 detail.

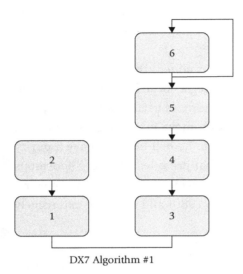

As a case study, let's look at Algorithm 1 (see Figure 5.5), which shows a total of six numbered blocks: two stacked on the left and four stacked on the right. The bottom blocks in all DX7 algorithms are carriers, those above are the modulators.

The left-hand column of blocks in Algorithm 1 is a straightforward modulator-to-carrier configuration. The right-hand column has three modulators: one fed by another modulator that is fed by yet *another* modulator. It is helpful to start at the bottom of the algorithm diagram with the carrier and work up when designing a sound. In fact, a great way to learn FM synthesis is to deconstruct preset patches you like by turning off all but one carrier, and step-by-step introducing each modifier and (if present) additional carrier(s) until you understand the purpose of each component.

Back to Algorithm 1 (Figure 5.4): Notice that the patch out from block 6 is not only feeding into block 5, it's also feeding back into itself! This is how noise is generated in FM. Noise, of course, is useful in many sound designs of percussive instruments or used as a layer, perhaps, with a short attack–decay envelope to act as the attack portion of a pitched sound.

Operators

Let's retire the term "blocks" at this point and replace it with Yamaha's term: *operator*. Each algorithm of the DX7 consists of six operators in thirty-two different configurations. Figure 5.6 shows the constituent parts of an operator: a sine-wave oscillator fed with pitch information from the MIDI controller (commonly your keyboard), a modulation input (used for some operators in a given algorithm), and an EG that is multiplied with the oscillator in a VCA.

Figure 5.6

DX7 Operator: Each DX7 algorithm has six operators in a variety of configurations to satisfy a wide range of timbral possibilities.

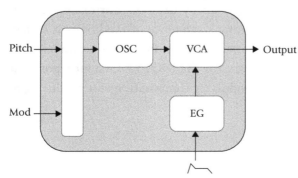

Why is there an EG in every operator? If the operator is used as a carrier, the EG is simply shaping its volume over time. If the EG is part of a modulator, it will shape the intensity of the modulation so that the complexity of the overtones will vary over time. Using EGs as much as possible in your operators will add dynamic interest to the patches you design.

> It is worth noting here that a discussion of filters may seem conspicuously absent in this chapter about FM. The harmonic complexity and control of the spectrum coming from the algorithms alone makes a filter somewhat redundant and unnecessary. Learning to craft a timbre by choosing the best algorithm option and shaping the operators accordingly will yield better results than applying a filter. That being said, some newer software instruments like NI's FM8 do offer LPFs with resonance control as an additional creative element.

Operators and the Harmonic Series

We already know that harmonic relationships between the carrier and modulator(s) are important for tonal sounds. In Chapter 2 we learned that the natural harmonic series is built upon integer multiples of the fundamental. It is therefore common to find carrier and modulator tuning controls offered in steps of the harmonic series so that harmonicity is the default. As options for the carrier and modulator, the Yamaha DX instruments provide thirty-two harmonics (fundamental plus thirty-one overtones), some newer instruments offer more. Some instruments configure operator frequencies so that normal tuning is done with ratios, whereas others like Logic's EFM1 have a dial for going through the steps of the harmonic series (see Figure 5.7). For detuning, instruments using ratios will accept decimals or small frequency offsets or have a fine-tuning control.

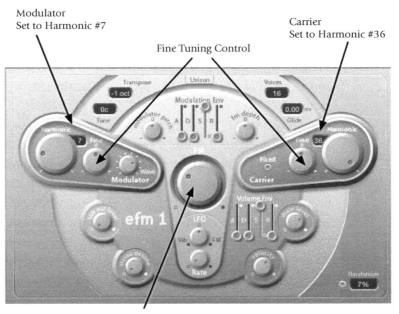

Figure 5.7

Logic EFM1 diagram.

Principles of FM Sound Design

Let's start making sounds with FM. If you have a hardware DX instrument or would like to use one, it is our recommendation that you begin with a plug-in that has a more inviting and logical graphical interface. The DX series' alphanumeric data-entry method can be confusing. Our favorite software-based FM synth is NI's FM8, which is built on a DX7 foundation but goes far beyond its capabilities, and the interface is relatively easy to understand. Let's use it to create some of the same sounds we designed in Chapter 4. Download a FM-capable synth, even if it's just a demo, to work your way through these examples.

Sound Design 5.1—Constructing Sawtooth Wave in FM8

There are a couple ways of estimating a sawtooth wave with FM synthesis (see Figure 5.8). Let's take them as Plans A and B.

Figure 5.8
Sound Design 5.1 sawtooth wave in NI's FM8.

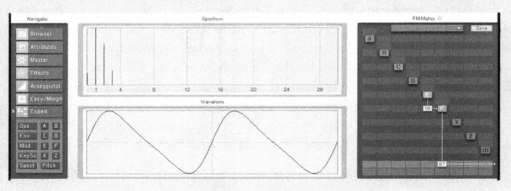

Steps for Plan A

1. Create a new, blank instrument. In FM8 only operator "F" will be connected. Playing a note will produce just the sine wave of the carrier.
2. Activate operator "E" by right-clicking on it with your mouse. Click and hold the light-gray box below E and left of F and drag it upwards to introduce the modulation signal to the carrier. You should start to hear harmonics added as you bring the level of E up. Set the level between 20% and 25% and you will hear something like a sawtooth wave (all of the harmonics are present but with lower amplitudes in the higher frequencies than you would get from a sawtooth oscillator).
3. By default the operators are set to the fundamental pitch (aka, the first harmonic), so in this configuration we are modulating with a frequency equal to the carrier, or a 1:1 ratio.

4. Play around with changing the amount of the modulating signal to get a sense for what aspect of the timbre you are controlling (for future reference).
5. Make a mental note of the fact that modulating a carrier with the same frequency results in generation of all harmonics.

The second way of creating a sawtooth wave is even simpler, and employs a technique covered later in the chapter: feedback (see Figure 5.9).

Figure 5.9
Sound Design 5.2 sawtooth wave using feedback instead.

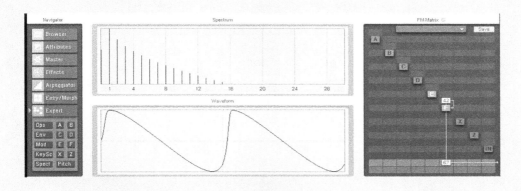

Steps for Plan B

1. With just one carrier (no need for a modulator), take its output and feed it back into its own modulation input.
2. With an input level of around 40% you should have a timbre that closely resembles that of a sawtooth!

Sound Design 5.2—Square Wave

Figure 5.10
Sound Design 5.3 square wave/clarinet-like sound using FM8.

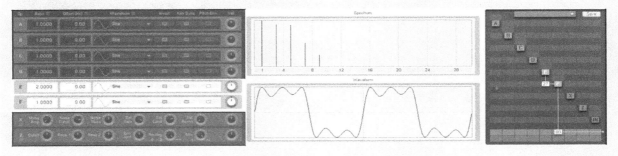

(Continued)

Sound Design 5.2—Continued

Continuing our theme of emulating sounds out of subtractive synthesis, let's try making a square wave (see Figure 5.10). We know that square waves consist of primarily odd harmonics, so working with a ratio that avoids even harmonics is the idea, and that can be done by having the modifier at a higher ratio than the carrier.

Steps

1. Create one carrier fed by one modulator.
2. Set the modulator's frequency to a ratio of 2:1 (so that it is an octave higher than the carrier).
3. Increase the modulation input on the carrier to around 30% to achieve a square wave.
4. Shaping the attack and release, and perhaps adding a little bit of noise to simulate breath/reed sounds, will resemble the timbre of a clarinet.

Sound Design 5.3—Reed Pipe Organ/Oboe Timbre using FM8

Setting the carrier to a higher frequency than the modulator results in a complex waveform containing odd and even harmonics, an extension of the sawtooth wave design in Sound Design 5.1. Note that the unused operators and controls are grayed out in the image to help you avoid confusion (see Figure 5.11).

Figure 5.11
Sound Design 5.4 reed pipe organ/oboe timbre using FM8.

(Continued)

Sound Design 5.3—Continued

Steps

1. Using the same patch as in Sound Design 5.2, click on "Ops" to bring up the operators (can be seen toward the bottom of the left-hand column in the "Navigator" section in Figure 5.9). You will see that E and F (the operators we have been working with) are both set to a 1.0000 ratio relative to the fundamental. For this design, set the carrier (F) to 4.0000. Remember, the first four steps of the harmonic series are (1) fundamental, (2) +1 octave, (3) up a fifth, and then (4) up a fourth, so the carrier is now two octaves higher than the modulator. If you are holding a C on the keyboard, that would be (1) C, (2) C an octave higher, (3) G, and then (4) C again, two octaves from where we started.
2. Setting the modulation level between 20% and 28% will result in the sound that (admittedly it takes a little imagination!) is oboe-like or a reed setting on a pipe organ.
3. As with Sound Design 5.1, explore lesser or greater modulation settings and see what sounds come through, and then play with different carrier ratios while leaving the modulator at 1.0000.

Modulation Index

Bear with this bit of math for just a moment:
The modulation itndex formula (see Figure 5.12) illustrates how the components of FM have an impact on the number of sidebands produced. A higher modulation index (β) value means more sidebands, a lower β means fewer.
  Details:

- β is the symbol representing the modulation index.
- ΔC in the numerator (top part of fraction) represents the carrier frequency's deviation. What that means is the distance in hertz from highest to lowest frequency covered by the carrier after being modulated. Of course, the deviation is a result of the modulator's amplitude because the greater that intensity, the wider the carrier's deviation. If that's confusing, think of the modulator's being a LFO resulting in vibrato instead: The amplitude of the LFO (or how much is being fed into modulating the VCO) is what determines the intensity of the vibrato.
- The denominator (bottom of the fraction) is the frequency of the modulator.
- A little carrier deviation (say, 500 Hz) with a modulator frequency would yield a β value of only 0.5. We can infer from this result that there are few sidebands present, and thus the timbre is mostly carrier.
- A βz value of 2 would suggest a carrier deviation of 500 Hz over a modulator frequency of only 250 Hz, resulting in more sidebands.

$$\beta = \frac{\Delta C}{M} \qquad \beta = \frac{\text{Carrier Deviation}}{\text{Modulator frequency}}$$

Figure 5.12
Calculating the modulation index.

Figure 5.13 Example of amplitude variation within sidebands.

Amplitude of Sidebands

Visible in Figure 5.13 is not just the presence of many sidebands but a clear variation in intensity. Predicting this mathematically gets deep into calculus with the Bessel function, well beyond the scope of this book. Suffice it to say, creative sound design with FM does not require the ability to solve complex formulas—although the math fans in the audience should have at it! Rather, studying cause and effect with the rules covered in this chapter will go a long way toward developing an intuition with the process that will be of greater value creatively.

Multiple Carriers

Using more than one carrier is simply a mixing process. Algorithm 32 on the DX7 is designed to function in an "additive," organ-like fashion. As you can see in Figure 5.14, operators 1–5 are just summed sine waves. In theory, each operator would be set to a different frequency within the harmonic series and balanced and shaped to achieve the desired timbre. Operator 6 can do the same, but notice that it also has a second output that feeds back into itself! (More on this in the next section.)

You can add depth through the slight detuning of one or more operators in a summing configuration such as this, in much the same manner as we did for the synth strings sound design in Chapter 4. When detuned just right, the result is a pleasing "phasiness" that adds depth and interest to a patch.

Noise, Distortion, and Feedback!

Noise is an important part of sound design, whether it be for articulating the beginning of a pitched sound or in creating a drum or percussion patch that is unpitched. So how do you get noise working with a bunch of sine-wave operators? The answer is using feedback, something alluded to in the description of Figure 5.14.

In the DX7's Algorithm 32, the sixth and final carrier in the sequence has a second out that feeds back into its own modulation input. At low levels the feedback introduces a distortion to the sound that can be described as gritty or brassy. As the level increases there is a point where the distortion takes on a square-/pulse-wave quality; pushed a little more it becomes metallic sounding. At extreme levels the feedback adds so many sidebands that the sense of pitch becomes nonexistent and you end up with noise. This is describing feedback into carriers, but you can also send modulators back into themselves as feedback, which begins as a thick, brassy, sawtooth quality and shifts into noise the higher the fed-back signal gets.

Feedback is a great creative tool in FM synthesis, and its possibilities should be tested and considered whenever you build a patch. It can be quite rewarding to experiment with

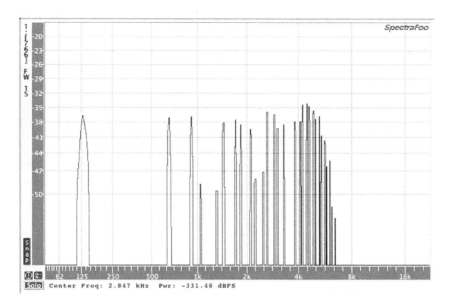

Figure 5.14
DX7 Algorithm 32.

feedback in different parts of a FM patch: carrier feeding back to its own modulator, modulator feeding back on its own modulator, and so forth. Be creative; what you hear may not be right for that particular need but may be something you can file away for future use. Software synths make it very easy to save patches, so anytime you get something interesting—even if it's not right for the project you are on at the time—save it as a patch! We recommend creating a folder (if your instrument allows) for interesting patches that you don't yet have a place for. Sometimes going through those when you need a bit of inspiration can be fruitful.

Multiple Modulators

Using more than one modulator applied to the same carrier is the key to building complex timbres. Perhaps the most straightforward consideration is using each of the modulators at different areas of the frequency spectrum in a quasi-additive type of sound design. In other words, a C4 (261-Hz) carrier might have a modulator working an octave or two higher and another at 3–4 octaves higher . . . and so on.

Key Sync/Key Tracking

Each operator will have the option to track the pitch coming from the MIDI controller or stay at a set frequency. Using key tracking will maintain the same timbre regardless of which range of the keyboard is being played. Table 5.2 suggests scenarios in which leaving key tracking off can be useful.

The Producer's Point of View on FM Synthesizers

There is no doubt that the use of FM synthesizers for music production had a huge impact on how electronic music evolved during the 1980s. In fact, the typical new "digital sound"

Table 5.2 Applications for Modulation (Key Tracking Off)

Frequency Range	Application
Low: <30 Hz	Will behave as vibrato; operating as a LFO
Medium: 100–500 Hz	May act as a formant, the resonant sound of a person's vocal cavity or the body of an instrument that is an important aspect of its timbre but independent of the pitch being played.
High: >1000 Hz	Adds color to the top end of the sound; brightness or articulation is accentuated but pitch independent.

that FM brought to the palette of music producers in the 1980s helped develop a whole new sound that carried through the 1990s, and even up to these days. As mentioned before, the DX7 was the poster boy for an affordable and relatively easy-to-use FM machine. Released in 1983, the original DX7 became quickly the must-have machine for any professional pop producer. Artists and producers such as Talking Heads, Brian Eno, Supertramp, Phil Collins, Depeche Mode, U2, The Cure, Toto, and many more took advantage of the new sonic potential that FM and the DX7 were offering. One of the biggest selling points of the DX7, when it first came out, was that it was one of the first affordable commercial synthesizers to have built-in MIDI ports. In fact, Andrea remembers when he went to buy a polyphonic synthesizer, right when the DX7 was launched; one of the biggest selling points advertised by the floor guy at the local music store was that it had "this new communication protocol called MIDI." The DX 7 eventually evolved into a series of very successful Yamaha sequels such as the much-improved DX7 II FD, the TX7 (a module version of the DX7), and the later TX-81Z and TX-802. Yamaha made also a few low-budget FM synthesizers such as the DX9, DX100, DX21, DX27 and the venerable FB01 (of which Andrea is still a proud owner!).

FM synthesis is particular effective when used to create sounds that are punchy and edgy with an underlying "metallic" quality that can be more or less pronounced depending on the producer's choice. Let's take a look at some of the most typical FM sonorities that made this type of synthesis legendary and that we can use for our own productions.

Typical FM Synthesis Patches and Sonic Characteristics

Because of its peculiar approach to generate more complex waveforms, FM features rich sounds that can be effective in both the low and the high end of the spectrum. In fact, basses, electric pianos, and bells constitute the core of the FM sound palette. Let's take a look at the main sonic features of FM in relation to its feature and main characteristics (Table 5.3).

FM reached its peak as "producer favorite" in the mid-1980s up to the mid-1990s in albums like Al Jarreau, "High Crime" (WEA,1984), Phil Collins, "No Jacket Required" (WEA, 1985), A-Ha, "Hunting High And Low" (Warner Bros.,1985), Michael Jackson, "Bad" (Epic,1987), Sting, "The Dream Of The Blue Turtles" (A&M,1985), Tangerine Dream, "Underwater Sunlight" (Jive Records, 1986), and many others. FM is still an incredible source for patches that can be used in a variety of situations and styles. Its double-faceted ability of creating celestial smooth electric pianos and aggressive and

Table 5.3 The Main Sonic Characteristics of FM Patches

Feature	Characteristics
Overall personality	FM patches have an overall personality that can be edgy and glassy. Its edginess can be either sharp or rough, depending on the type of algorithm or operator used. In general, FM patches have much more aggressive personalities than the ones created on subtractive synthesizers.
Edginess/smoothness	FM presents some definite edgy and punchy traits that come from its ability to generate waveforms with a high harmonic content.
Transparency	FM can be very transparent when used in the higher register and with high ratios between oscillators.
Thickness	It is possible to obtain a good variety of thicknesses with FM. Depending on the matrix used we can have heavy sonorities (as in some classic FM bass patches) or light and thin electric piano, bell, and mallet instruments.
Punch	FM is capable of delivering really punchy patches such as synth basses and Clavinet-like keyboards. In particular the synth basses of the DX7 have become legendary when used by artists such as Madonna ("True Blue," Warner Bros., 1986) and Michael Jackson ("Bad," Epic, 1987).
Weight	FM's weight is, in general, lighter when compared with other types of synthesis. The overall weight of a patch can be increased by using a matrix with a higher number of operators.

punchy basses allows us to use it in very diverse situations. One of our favorite uses is on electronic dance music (EDM) tracks where we need aggressive and edgy sonorities. In particular, you can use it in styles such as House and Dubstep.

FM Synthesis Hardware and Software Sources and Options

There are several options these days when it comes to integrating FM synthesis in your project studio setup. Although the hardware options rely mainly (but not only) on vintage gear, the software options are many and keep constantly expanding. These days, for FM patches the software synth options provide much more flexibility and ease of use than their hardware counterparts. This is particular true for vintage FM synthesizers. Early devices, such as the DX7, were pretty hard to program and limited in terms of options. No matter which path you decide to take, it is really important for you to have at least one or two solid FM synthesizers in your studio. Let's take a look at some of the most used FM synthesizers available to the modern electronic producer (Tables 5.4, 5.5, and 5.6,).

The number of FM and hybrid software synthesizers that are available is increasing exponentially, and several are also available for iPad and Android Tablets. If you are new to FM synthesis we recommend getting one of the soft synths for iPad

Creating Sounds from Scratch

Table 5.4 Selected Vintage Hardware FM Synthesizers

Vintage Hardware FM Synthesizers	Features
DX7 Yamaha	This is the original affordable FM synthesizer that Yamaha brought to the market in 1983. It was a truly revolutionary machine, not only for its "relatively affordable" price, but also because it had built-in MIDI ports.
DX7 mkII DX1 DX5	These are evolutions of the original DX7. The mkII was a much more powerful machine with a 16-bit engine, double the original polyphony, enhanced MIDI support, microtuning capabilities, aftertouch-controlled pitch bending, and multiple LFOs.
TX7 TX-802 TX-81Z	The TX series are basically sound module versions of the DX synthesizers. The TX-802, a rack-module version of the DX7mkII, was the flagship machine capable of 8 parts multi-timbral operation for sequencing.

because they are really cheap but offer a nice view over the great possibilities of FM (see Figures 5.15 and 5.16).

Sonic Categories and FM Patches

As was the case for subtractive synthesis, FM is particularly suited for certain types of patches that can be effectively used in our productions. The complexity and sophistication of FM make it a versatile tool for creating a larger variety of sounds compared with that of its distant "cousin," the subtractive approach. For this reason we usually like FM for patches that are more refined, a bit more stylish, creative, and sophisticated. Among our favorite categories of FM sounds are the basses, the electric pianos, mallets (synth and semi-acoustic), pads, and organs. Let's take a look at these categories and identify what makes them so special.

Table 5.5 Selected New Hardware FM Synthesizers

New Hardware FM Synthesizers	Features
Nord Lead A1	The A1 combines several synthesis techniques to produce rich and complex patches. Its oscillators can produce up to forty-seven different waveforms. OSC 1 can be modulated by a sine wave generated by OSC 2.
Kronos X, Korg	The highly sophisticated Kronos by Korg combines several synthesis techniques that can be easily programmed through its large graphic display. Its software-based engine allows you to use not only FM but also subtractive, wave shaping, physical modelling, and many others.

Table 5.6 Selected FM Software Synthesizers

Software FM Synthesizers	Features
FM 8, Native Instruments	FM 7 first and FM 8 later are considered the real descendents of the original DX7. In our opinion FM 8 is a tool that should be present in any studio. Its rich sound palette, easy-to-use interface, and extremely flexible and powerful engine make this software synthesizer a must-have!
Tassman, Applied Acoustics Systems	Tassman by AAS is completely modular system that allows you to modulate the carrier in any way you want. Although a bit hard to program, it offers almost infinite options when it comes to original FM (and hybrid) sound creation.
Logic Pro X, EFM1	This software synthesizer is bundled with Logic Pro X. Although it is pretty simple, it is a great tool for understanding how FM synthesis works. Even if it is simple, EFM1 is capable of generating some really interesting and original sonorities.
Operator, Ableton	Operator is excellent if you use Ableton Live. The amount of customization is great (you can import or design your own waveforms) and the simple interface makes it fairly easy to program. A must-have if you use Live.
Sytrus, Image Line	Sytrus is an excellent software synthesizer that is capable of much more than just FM synthesis.
DXi, Takashi Mizuhiki (Figure 5.15)	This soft synth for iPad is a fantastic choice for learning how to program in FM. It features four oscillators combined in several algorithms that are easy accessible. You can choose among twelve waveforms and the interface reminds you a bit of the original DX7.
TF7, Pier Lim (Figure 5.16)	TF7 is another soft synth for iPad that is incredibly powerful. It features seven starting algorithms (expandable via an app purchase) and a great selection of controllers to create some unique FM sounds.

Bass

The FM bass patches have literally defined an era of electronic music. It is one of our favorite synthesized bass sonorities. Artists like Madonna, Michael Jackson, and Tears for Fears have made this sound legendary (see the subsequent Listening Corner). Its main characteristic is based on a dry, punchy, and "in-your-face" sonority. Depending on the algorithm used, it can span from a "woody" timbre all the way to a more metallic character. Listen to Audio Examples 5.1–5.4 for some examples of FM basses. An FM bass patch allows you to create parts that are more complex and rhythmically active than a typical subtractive one. This feature usually calls for snappier and "jumpier" lines such the one shown in Figures 5.17 and 5.18.

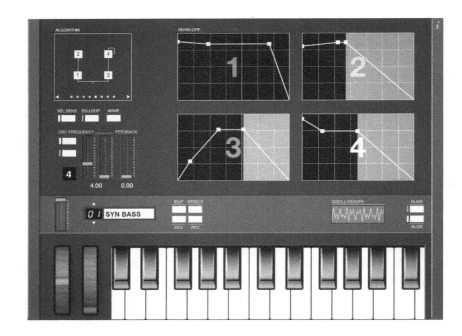

Figure 5.15
DXi, FM synthesizer for the iPad.

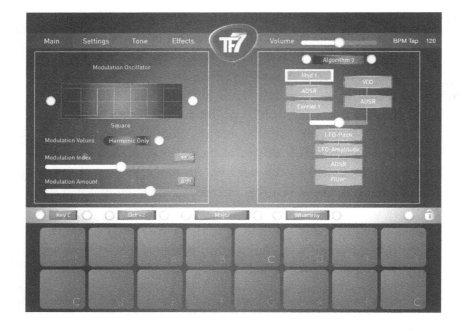

Figure 5.16
TF7, FM synthesizer for the iPad.

Figure 5.17
An example of a bass line for a typical FM bass patch.

Figure 5.18
Another example of bass line for a typical FM bass patch.

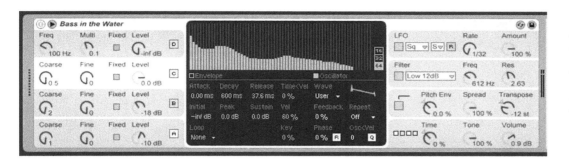

Figure 5.19
An example of an aggressive bass patch created in Operator with a customized oscillator waveform.

For genres like Dub Step, for which you need a bass with more body, an aggressive attitude, and wide harmonic content, FM provides the sounds you need. In the patch shown in Figure 5.19 we used Operator in Ableton Live to create an aggressive bass patch that is well suited for a Dub Step "wobbly" bass. There are four oscillators here that are connected in series. For OSC C we used a customized waveform that we drew directly in Operator.

To create the legendary "wobbly" effects on the bass patch, we have assigned the LFO (with a square wave) to control the cutoff frequency of the filter (Figure 5.20), and the rate of the LFO is controlled in real time by the Modulation wheel (CC#1), as shown in Figure 5.21.

The final effect is that the LFO, via its square wave, moves the frequency of a LPF with a 12-dB slope, which has the effect of opening and closing the sound; the speed of the filter is controlled in real time through CC#1 - Mod Wheel. Look at Figure 5.22 for a diagram of how the filter of this patch is controlled by the LFO and the Modulation wheel, and listen to Audio Example 5.5 to hear the effect.

As usual, feel free to experiments with your sonorities. Sometimes in FM synthesis is a bit hard to predict exactly how a sound will turn out. This is true particularly when different algorithms and operators are used. Our suggestion is to start simple, with just

Figure 5.20
The LFO assigned to the cutoff frequency of the filter with a square wave.

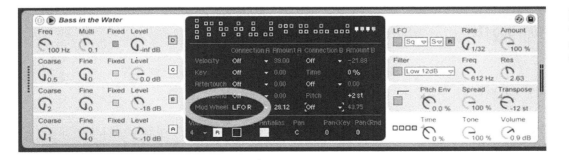

Figure 5.21
The modulation wheel assigned to the LFO's rate.

Creating Sounds from Scratch

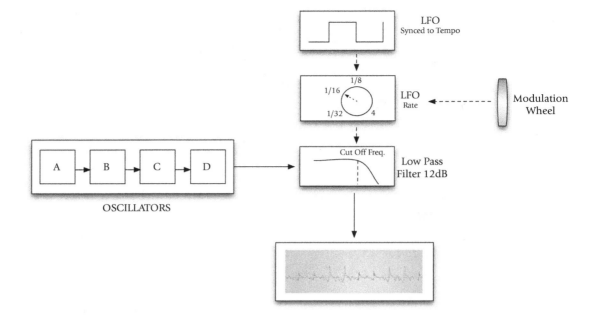

Figure 5.22 Diagram of a filter controlled by a LFO and the modulation ("mod" wheel).

a carrier and a modulator and then start adding other oscillators. Every time you add a new one use the "amount" to detect what that particular new waveform adds to the mix.

Mixing the FM Bass

When it comes to mixing the FM bass patches, one thing to keep in mind is their high harmonic content and their sharp attack. These are important sonic characteristics to make sure are enhanced during mixing. Use an EQ to try to enhance some of the key frequencies of your bass patch (see Table 5.7).

Table 5.7 Key Equalization Frequencies That Apply to FM Bass Patches

Frequencies	Application	Comments
60–95 Hz	Boost to add fullness	This range gives you a fuller sonority without overaccenting muddiness.
200–300 Hz	Cut to reduce muddiness	• Use with discretion. By cutting too much you might reduce the fullness of the bass track; by boosting too much you might increase its muddiness. • Use with a high Q value in order to be more selective.
400–600 Hz	Boost to add presence and clarity to bass	
1.4–1.5 kHz	Boost for intelligibility	
4–5 kHz	Boost for extra "slap"	Boost this range to add an extra punchiness and slap to traditional FM basses

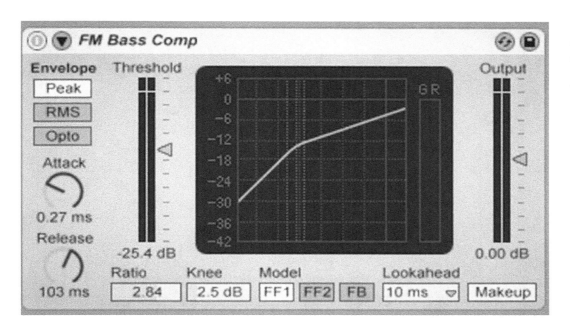

Figure 5.23
Compression settings for a FM bass patch in Ableton Live.

If you are looking to add extra punch, add a compressor with mild settings, with a short attack and a ratio between 2:1 and 3:1 (Figure 5.23).

To add a bit more character and to have the bass sit better in the mix with the rest of the keyboards we like to add a bit of chorus. This should be done with a gentle touch and not too dramatically. Usually a delay with a HPF around 150 Hz and a rate between 1 and 1.2 kHz will do the trick (Figure 5.24).

Listen to Audio Examples 5.6 and 5.7 to compare an FM bass with and without effects respectively (EQ, compression, and chorus).

As was the case for the subtractive bass, we recommend panning the bass in the middle of the stereo image.

For some classic example of FM basses, consult the listening corner in Table 5.8.

Electric Pianos

An electric piano is another signature sonority of FM synthesis. As we saw for the bass, the versatility of this type of synthesis allows us to create great sounds that range from

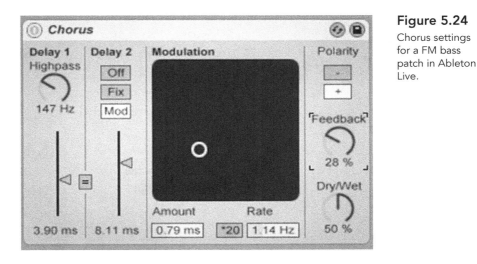

Figure 5.24
Chorus settings for a FM bass patch in Ableton Live.

Table 5.8 Listening Corner: FM Bass

Artists/Track/Genre	Featured Bass Lines
Michael Jackson, "Another Part of Me," from "Bad," Epic, 1987, Pop	This is a classic FM bass patch from the DX7. It is punchy and sharp with a metallic sonority underneath.
Madonna, "Who's That Girl," Sire Records, 1987, Pop	The typical sharp DX7 Bass
Tears For Fears, "Shout," from "Songs From The Big Chair," Mercury, 1985, Pop	Again a perfect digital FM bass from the legendary DX7

silky and smooth electric piano patches to more aggressive and edgy ones that cross into electric Wurlitzer or Clavinet territory (see Table 5.9 and Figures 5.25–5.27). FM electric pianos are ideal for pop ballads, but they can be effectively used also for more "loungy" styles like those of Ambient, Acid Ja, and Nu Ja.

Mixing the FM Electric Piano

FM electric pianos are a lot of fun when it comes to mixing. They can be treated in a million ways, and they always give good results. One of the key effects that we recommend

Table 5.9 Key Elements of FM EP patches

Type of FM EP Patch	Key Elements	Comments
Light and thin EP	• Built on three sets of modulator/carrier with sine waves (Figure 5.11) • Fairly well-distributed harmonic content with emphasis on mid–low and high frequencies	This is the typical FM EP that features a nice clear tone that cuts easily through the mix.
FM Wurlitzer	• It features a "rougher" and more distorted sound with richer harmonics (Figure 5.12) • The oscillators modulate in groups of three in order to get a more interesting and aggressive sound	A great way to get a "dirtier" EP that can fit in any styles where sharper attack and higher richness are needed
FM Rhodes	• It features a mellower and muffled sound with a lower level of harmonics (Figure 5.13)	The typical Rhodes sound, with a slightly more "analog" personality, not as bright or aggressive as the previous patches

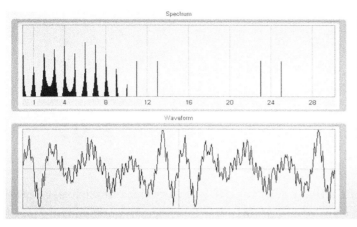 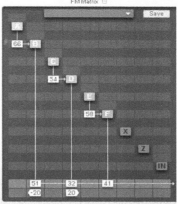

Figure 5.25
Spectrum and waveform of a clear electric piano patch in FM8.

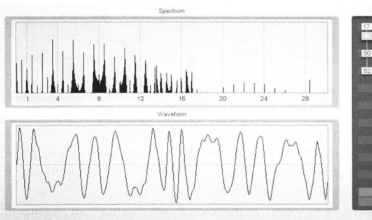

Figure 5.26
Spectrum and waveform of a Wurlitzer patch in FM8.

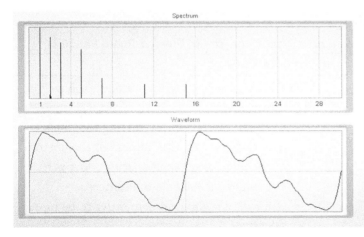 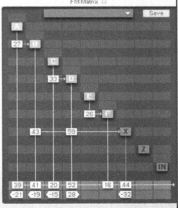

Figure 5.27
Spectrum and waveform of a Rhodes patch in FM8.

using is a nice and lush chorus. In fact, this modulation effect has become part of the distinctive sonority of these patches. The chorus adds depth and substance to the patch, making it less sterile and giving it more personality. To improve the intelligibility and to give more space to other instruments in your mix we also like to spread the electric piano patch on the stereo image by adding a slight stereo tremolo synced to the tempo of the piece (Figure 5.28).

Creating Sounds from Scratch

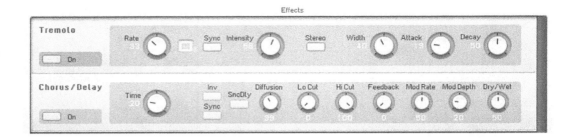

Figure 5.28
Chorus and tremolo effects assigned to an EP patch in FM8.

This creates a nice stereo effect that makes the patch bigger yet not intrusive. Be careful not to overdo it unless you want to create some special effects. A good starting point is a medium–low width (the spread between the left and the right channel) and medium intensity. Listen to Audio Examples 5.8 and 5.9 to compare an electric piano without and with chorus and tremolo effects.

If you want to add more byte to the patch you can add a subtle saturation via a Tube Amp simulator. Just add a bit of drive without going overboard; your sound will be more present and aggressive.

Applying a gentle EQ can be very effective to bring out some of the key frequencies of the FM EP. Refer to Table 5.10 for a list of the important frequency ranges.

The addition of a nice medium to large reverb has been the signature of some of the best FM pianos in the 1980s and 1990s. Nowadays we recommend using little reverb on these pianos in order to minimize the "retro" effect. Usually a 1.6- to 1.8-s reverb will do a nice job when you are trying to place the instrument in a real space.

For some classic example of FM electric pianos, consult the listening corner in Table 5.11

Table 5.10 Key Equalization Frequencies That Apply to FM Electric Piano Patches

Frequencies	**Application**	**Comments**
60–95 Hz	• Boost to add fullness	• Use this range to add more body and weight to the patch
200–300 Hz	• Cut to reduce muddiness	• Use with discretion. By cutting too much you might reduce the fullness of the bottom of the piano track; by boosting too much you might increase its muddiness. • Use with a high Q value in order to be more selective.
400–600 Hz	• Cut to reduce nasal sonorities	
1–2 kHz	• Boost for presence • Cut for leaving space for vocals	
8–9 kHz	• Boost to add more metallic sonorities to the patch	Boosting this range adds some high-end metallic sonorities

Table 5.11 Listening Corner: FM Electric Pianos

Artists/Track/Genre	Featured Bass Lines
Phil Collins, "Take Me Home," from "No Jacket Required," Atlantic Records 81240-1, 1985, Melodic Pop	This is a subtle EP with a thin timber that highlights mainly the most bell-like timber of the DX7.
Simply Red, "Holding Back the Years," WEA – 960 452-1, 1985, R&B Pop	A nice mellow electric piano playing a simple and slightly syncopated quarter note pattern.
Al Jarreau, "Love Speaks Louder Than Words," from "High Crime," Warner Bros. Records – 1-25106, 1984, Soul Pop	A more aggressive electric piano that borderlines with a synth pad probably played on a DX-1.

Mallets

Another key signature of FM synthesis is synth mallet sounds. These can be very versatile, ranging from some realistic edgy vibraphones, to warm marimba, all the way to more electronic and metallic sonorities. The creation and modification of these patches is fairly easy and fun to do. The main sonic characteristic of FM mallets is the ability to be able to cut through the mix but at the same time, if necessary, provide deep and woody low frequencies that help filling the low end of your mix. Let's take a look at some of the main FM mallets' sonic features (see Figures 5.29–5.31).

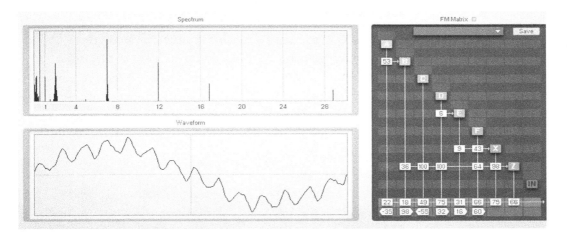

Figure 5.29
Spectrum and waveform of a typical FM vibraphone patch in FM8.

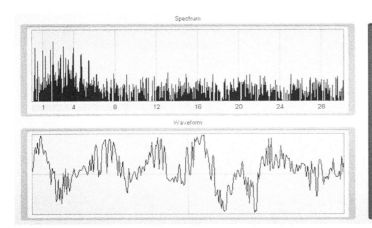
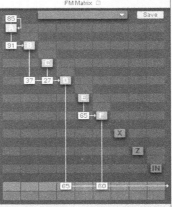

Figure 5.30
Spectrum and waveform of a typical FM marimba patch in FM8.

Creating Sounds from Scratch

Figure 5.31
Spectrum and waveform of a typical FM glockenspiel patch in FM8.

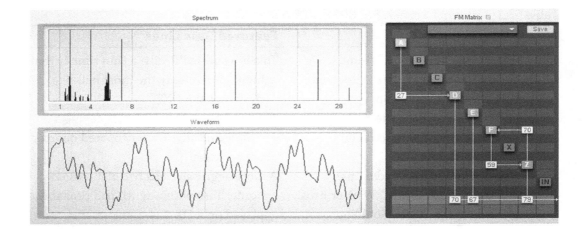

Mixing the FM Mallets

Mallet sounds usually do not have a lot of flexibility when it comes to equalization. In general we recommend using the same suggestions we gave for the electric piano because both types of patches share a similar range (Table 5.12).

Table 5.12 Key Elements of FM Mallet Patches

Type of FM Mallet Patch	Key Elements	Comments
FM vibraphone	• Built on a complex matrix of modulator/carrier with sine waves (Figure 5.29) • Harmonic content is concentrated on low harmonics with the addition of some very high ones (Figure 5.29)	This is a typical vibraphone patch that has a very natural and realistic character. If you need to add a bit more byte and shine, you can change some of the waveforms to square.
FM marimba	• This patch is characterized by a very rich harmonic content (Figure 5.30) • The oscillators (sine waves) are arranged in a more complex matrix system (Figue 5.30)	The woody sound is provided by the lower harmonics, and the mid and highs ones allow you to give a more vibrant ringing metallic color. If you use more complex waveforms for the modulators you can obtain sonorities that get closer to those of a guitar
FM glockenspiel	It features a thinner and more bell-like sound (Figure 5.31) Strong harmonics are featured in the mid–low area with steady descending intensities for harmonics in the mid and high spectra (Figure 5.31)	This is a typical example of FM creating some interesting and realistic bell sounds.

In terms of panning we like to pan the mallets on a fairly wide stereo image with the low range of the instrument to the left and the high range to the right of the stereo image. Not all FM synthesizers allow you to adjust the stereo width based on the range though. To give the illusion of a mellower and more liquid sound we like to add a light chorus on a mallet patch that has more of a metallic flavor. This will allow you to take off some of the harshness and edginess of such a patch. Be careful not to overdo it though because it could take away too much presence.

To add some nice ambience we like to use a medium reverb with a decay time between 1.6 and 1.8 s and some good predelay in order to preserve the attack of the patch (Figure 5.32).

FM mallet sounds are very effective when used for fast passages with alternation of notes with low and high velocities that bring out the contrast between mellow and edgier tonalities. For this purpose we like assigning a good arpeggiator to a FM mallet patch. FM8 has a built-in arpeggiator with some really excellent features. When using this option, make sure that you accent some notes in the pattern in order to bring out a brighter color now and then (Figure 5.33).

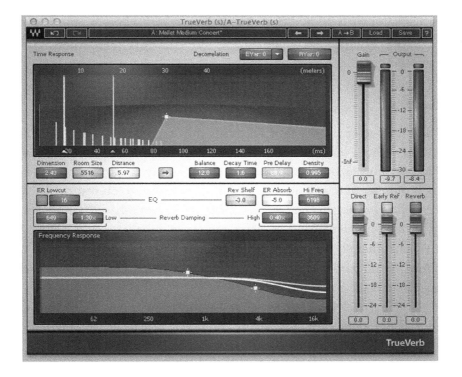

Figure 5.32
A good starting point for a natural reverberation on a mallet patch.

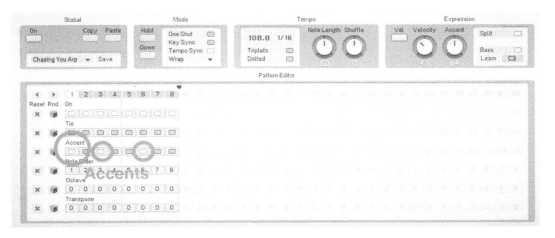

Figure 5.33
An example of a FM patch triggered by an arpeggiator in FM8.

Table 5.13 Listening Corner: FM Mallet Patches

Artist(s)/Track/Genre	Featured Bass Lines
Sting, "Love Is The Seventh Wave," from "The Dream Of The Blue Turtles," A&M, CD 3750, 1985, Reggae Pop	This is a good example of FM marimba with a metallic accent. The sharp attack and the short release are typical of this sonority.
Toto, "Africa," from "Toto IV," Columbia, FC 37728, 1982, Rock Pop	The kalimba sound in the intro is a sound created on a precursor of the DX7 called GS1.
Harold Faltermeyer,. "Axel F," from "Beverly Hills Cop Soundtrack," MCA Records, MCAD-5553, 1985, Sound Track	The nice marimba and bells sounds come in at around 1:44, a great example of how elegant and classic FM sounds can be. The light reverb and delay add a nice environment to the patch.

Listen to Audio Example 5.10 for an example of a FM mallet patch triggered by a FM8 arpeggiator with the addition of a medium reverb, chorus, a touch of delay, and phaser.

For some classic example of FM mallets patches check out the listening corner in Table 5.13.

Pads

Pads, no matter what type of synthesis you use, can always be a great area for experimentation. Usually you can get great results with any type of synthesizer and/or synthesis. FM is not an exception. Because of its capability to generate complex waveforms and a high number of harmonics, we can create some very interesting patches. Overall we like FM pads that retain the typical "glassy" and crystal-based sonorities. These patches are great for complementing an electric piano or a mallet sound, adding more substance to the instruments, and outlining the harmonic progression. Rarely we would use them for a foreground part. Instead they are great for being in the middle or background, gluing all the rhythm section together.

In Figure 5.34 you can see a typical spectrum graph of a FM pad with a simple waveform generated by only sine waves and featuring a low harmonic content (Figure 5.35).

Figure 5.34

The spectrum graph of a simple FM pad in FM8.

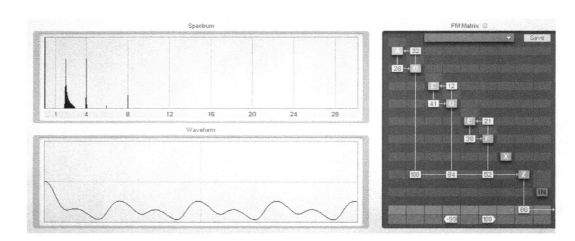

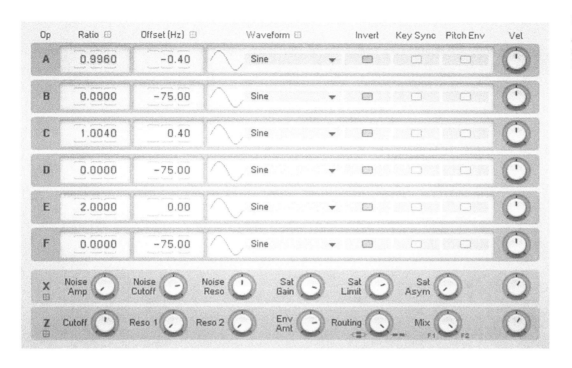

Figure 5.35
An example of a simple FM pad in FM8.

Listen to Audio Example 5.11 for a sample of a nice light FM pad, as shown in Figures 5.34 and 5.35.

If you are looking for something more aggressive you change the waveform types of the first oscillators. Usually going to a triangle or square wave for the modulators will add more character to the patch (Figure 5.36).

Listen to Audio Example 5.12 for a sample of a more aggressive and complex FM pad, as shown in Figure 5.36.

As it is the case for most of the pad patches, we like to have a fairly long attack and medium/long release. These settings will give you the necessary smoothness in order to have a patch that is not too intrusive and that can let the attack of the other parts (such as electric pianos, mallets, guitars, etc.) cut through the mix. Feel free to experiment though, because you can create some nice hybrid electric piano /pad patches by using a shorter attack and brighter filter settings, as you can hear in Audio Example 5.13, where we have shortened the attack of the oscillators of Audio Example 5.10.

Figure 5.36
An example of a more complex FM pad in FM8.

Creating Sounds from Scratch

Mixing the FM Pads

When it comes to pads you have a lot of freedom in terms of mixing. A medium-to-long reverb helps to move the pad farther from the forefront of the mix, which usually helps in blending it with the rest of the instruments. We would start with a length of 2.1–2.3 s. For FM pads we like to use a warmer reverb. A good synthesized algorithm can add a bit of warmth to the patch without adding muddiness. Modulation effects such as chorus and flanger can add a nice depth and character to a FM pad, bringing out some nice nuances created by the richness of the harmonic content. If possible, try to spread the pad over the stereo image; this will add character but also will allow an easier integration of the pad with the rest of the instruments.

Listen to Audio Example 5.14 for a sample of a smooth FM pad with the addition of chorus, flanger, and reverb.

Organs

The FM-based organs are among some of our favorite patches for this type of synthesis, along with the electric pianos. The wide range of variations that is possible makes this category a constant source of new patches and colors. FM can create convincing and inspiring organ-based patches ranging from classic pipe organs, to inspiring electric organs, all the way to accordions and electronic organ-like pads. A basic traditional organ pipe can be easily recreated by use of all sine waves for the oscillators (Figure 5.37).

Notice how in Figure 5.37 the ratio between oscillators is mainly integer based, with the exception of two oscillators (C and F). These two detuned frequencies add nice extra color and natural effect that is typical of pipe instruments. Listen to Audio Examples 5.15 and 5.16 to compare a pipe organ patch without the detuned oscillators and with the detuned oscillators, respectively. Notice how the additional oscillators with a noninteger ratio add higher harmonics to the spectrum (Figure 5.38).

To obtain a bigger church organ effect, for which the number of harmonics is higher and the sound is richer, we can use a more complex matrix, where three sets of modulators–carriers are spread over the stereo image. Listen to Audio Example 5.17 for an example of a bigger church organ.

Figure 5.37 An example of a classic pipe organ in FM8.

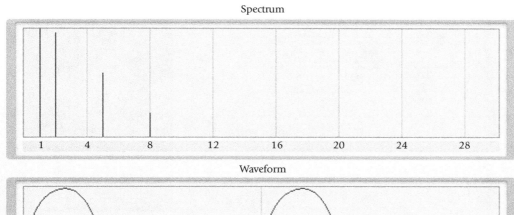
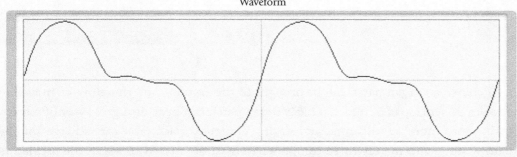

Figure 5.38
Spectrum comparison between a pipe organ patch without (upper) and with (lower) oscillators with noninteger ratios.

To create the effect of a more electro-organ such as the B3 with a nice percussive attack that is characteristic of the original instrument, we can use a more sophisticated combination of waveforms for the modulators (Figure 5.39).

In Figure 5.39 you can see how oscillators C, E, and F form the main body of the patch with ratios set at 0.49, 2, and 1, using some more complex waveforms. What gives that great percussive sound, though, is the modulator A with a ratio of 7 that modifies carriers D and F, spread over the stereo image. Listen to Audio Example 5.18 for a sample of the FM electric organ shown in Figure 5.39.

Creating Sounds from Scratch

Figure 5.39
A FM electric organ patch.

Of course an organ patch can be brought to the extreme and therefore be moved in the realm of synth pads. In fact, a basic organ patch can provide a great way of experimenting into more adventurous and daring electronic pads. You can achieve this by experimenting with mainly three components: first, the matrix in which the oscillators are organized. In Audio Example 5.19, for example, we created a brand new patch starting from a traditional electric organ patch but simply changing the FM matrix (Figure 5.40).

Notice how, in order to create a more complex final waveform, the oscillators keep modulating each other from A through F with a double modulation occurring between A–B and C. This creates a very rich set of harmonics at every level (Figure 5.41).

Figure 5.40
The FM matrix of a synth pad based on an electric organ patch.

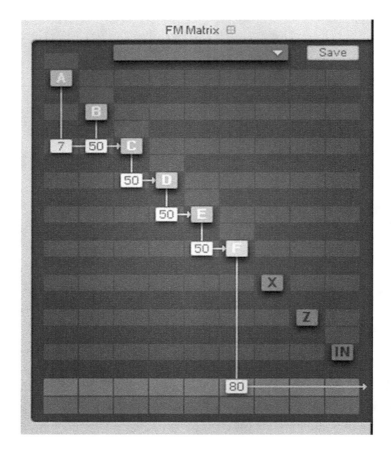

Frequency Modulation Synthesis

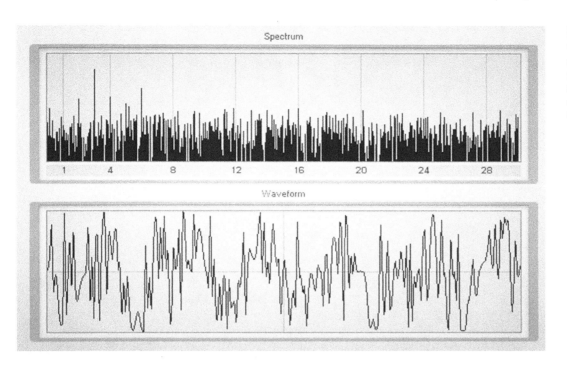

Figure 5.41
The spectrum analysis and waveform of a synth pad based on an electric organ patch.

The "cascade" modulation allows us to create more aggressive and edgier final waveforms. To give it more of pad feel we also add a slower attack and release. This allows us to reduce the "byte" of the patch, making it a bit smoother and more "dreamy" (Figure 5.42).

To add a more synthesized character to the patch you can also modify some of the waveforms of the oscillators. In this case (Figure 5.43) we have modified oscillators A and F with more interesting waves.

The final touch for a successful pad is the addition of some effects. In particular, we find that chorus, delay, reverb, and tremolo can be very helpful in adding a more

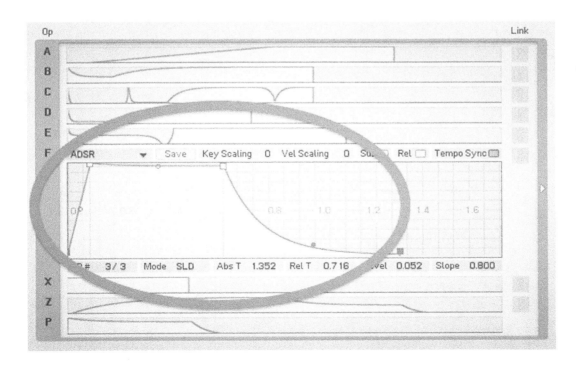

Figure 5.42
Envelope of the amplifier for the final oscillator F.

Creating Sounds from Scratch

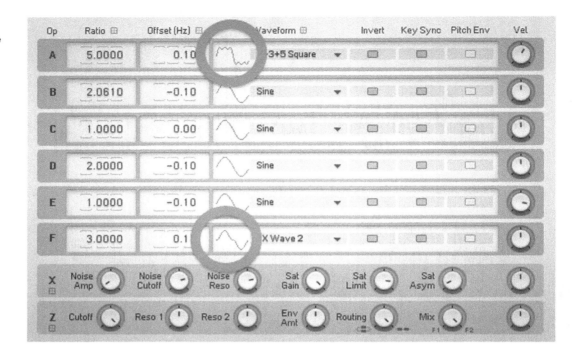

Figure 5.43
Waveforms of the oscillators of a synth pad patch based on an electric organ.

"pad-like" character to a FM patch. Listen to Audio Example 5.20. This is the same patch as that of Audio Example 5.19 but with the addition of a tremolo, a delay, and a chorus/delay (Figure 5.44)

The tremolo effect allows us to add a subtle movement across the stereo image that makes the patch more alive and less boring. The delay adds a sense of majesty and power to it, almost a feeling of ancient times. The chorus helps smooth the edginess of the original waveform.

This is just an example of how you can use FM for things that are more creative than just the typical sonorities. We think it is important to master the sounds that FM is better know for and then take those as a starting point for more daring and creative original patches.

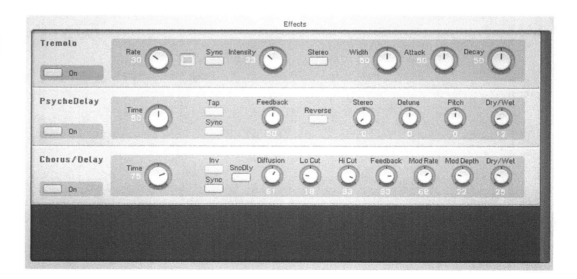

Figure 5.44
Effects applied to a synth pad patch based on an electric organ.

Exercises

5.1 Program two types of FM electric basses based on the following specifications:

Bass 1, edgy and sharp

Bass 2, aggressive

5.2 Program a silky and warm FM electric piano patch.

5.3 Using the patches created in Exercises 5.1 and 5.2 sequence a 16-bar riff, making sure to appropriately quantize the parts.

5.4 Program two FM mallets patches based on the following specifications:

Vibraphone, edgy and metallic

Marimba, "woody" and fat

5.5 Using the sequence recorded in Exercise 5.3 as a starting point, add a rhythmic mallet part.

5.6 Program two FM pad patches based on the following specifications:

Pad 1, smooth and light

Pad 2, aggressive and edgy

5.7 Using the sequence recorded in Exercise 5.5 as a starting point, add a pad part.

6

Additive Synthesis

Introduction

Additive Synthesis can best be described as sound design through manual assembly of sine waves starting with the fundamental and then the overtones necessary for achieving the desired timbre. Earlier we learned that through Fourier analysis it is apparent that sine waves are the basic building blocks of all complex sounds. Pitched sounds have harmonics that follow integer multiples of the fundamental frequency; nonpitched sounds simply have a fundamental with *in*harmonic (noninteger multiple) overtones. Armed with this information, Additive Synthesis offers a level of customization and control unequaled in techniques that predate it.

Additive Synthesis works on the premise that if we have a lot of sine waves at our disposal we can recreate most any sound by effectively reverse-engineering Fourier analysis. In practice, that is easier said than done but it does offer design possibilities that are not readily available in other forms of synthesis. Subtractive synthesis, for example, relies on coarse tools for shaping a sound's spectrum; FM synthesis offers dramatic timbral possibilities through its various combinations of operators; but detailed, manual control over individual harmonics is the province of Additive Synthesis.

Arguably the first implementation of additive sound design was with the pipe organ. Pipes of varying lengths are arranged in a harmonic series and activated by pulling *stops* on the organ console—thus allowing the organist to customize within a timbral range from thin, sparse tones to ones that are rich and dense. Of course, in this example the pipes were not producing pure sine waves—far from it—but the concept of building sound this way predates synthesizers *and* electricity.[1]

The Hammond B3 uses spinning tonewheels (see Chapter 1 for more details) to generate something closer to a sine wave. Rather than using stops, the Hammond used *drawbars* (see Figure 6.1) to actuate the output level of the tonewheels and enabling the assembly of up to nine harmonics.

The organ was a big step as a precursor to synthesizers but lacked a significant component: a means for shaping the volume of *individual* harmonics over time, an envelope generator (EG). True Additive Synthesis requires many EGs— in fact, it is ideal and typical to have a dedicated EG for *every* harmonic.

---

[1] The first pipe organs, which go back as far as the third century BCE, were entirely mechanical, no electricity was needed.

Figure 6.1

The drawbars on a Hammond B3 organ (Logic's Vintage B3 organ shown) are arranged in an additive fashion with each representing a harmonic in the overtone series. The first two drawbars are unused in this image because they break the pattern a little: The lowest is labeled 16′ and is an octave below the first white drawbar and the second is a fifth above the first drawbar. Engaging just the 16′ results in a lower, suboctave being added, whereas both together induce a psychoacoustical phenomenon known as *the missing fundamental*. When our hearing system detects a low perfect-fifth interval (as the first two drawbars would), our brain synthesizes a tone an octave below that: In other words, if harmonics 2 and 3 are present, a lower fundamental must be there somewhere! A cool trick is also used in plug-ins like MaxxBass from Waves, which synthesize a low fifth from a fundamental in the mix so that even when listening on small speakers your brain makes you think you are hearing a fundamental at an octave lower.

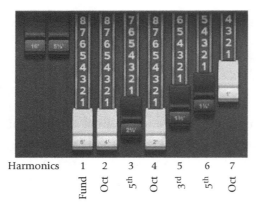

Figure 6.2 offers three views of a single note captured from a grand piano, each helping us dissect the complexity of the sound by telling different parts of its story. Figure 6.2a shows a spectrum analysis of the peak intensity of the harmonic content, but only at a single a moment in time (the attack is shown in this example). The fast Fourier transform (FFT) images in Figures 6.2b and 6.2c help us visualize how the harmonic structure of the note shifts dynamically over time. Figure 6.2b shows a traditional "waterfall" display in which the back of the three-dimensional (3D) image represents the start of the note, and the part closest to our perspective is the release; Figure 6.2c is a different kind of 3D FFT view, using color variation to represent intensity (seen here reprinted in grayscale; only frequency and duration are shown).

It is evident from Figure 6.2 that to reproduce something close to the piano timbre we are going to need a *lot* of sine waves. The fundamental frequency of D#2 is 77.78 Hz, and some level of each harmonic is present up at least through the audible spectrum: 155.56,

Figure 6.2
FFT of D#2 played on a grand piano: (a) peak spectrum analysis, (b) "Waterfall" FFT, and (c) spectragram.

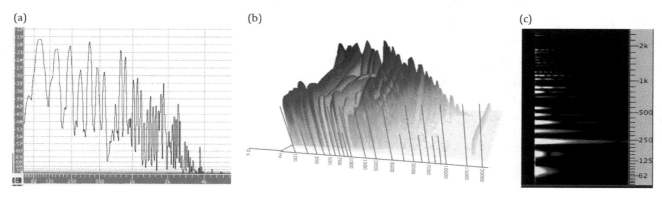

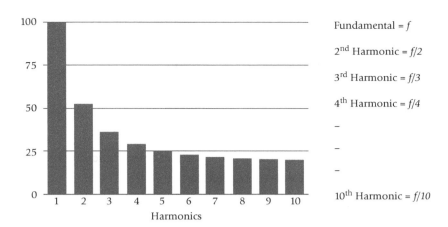

Figure 6.3
A sawtooth wave is made up of all harmonics descending in amplitude at a rate of roughly 1/n, where n is the harmonic number 1, 2, 3, etc.).

233.34, 311.12, and so forth. An analysis of the harmonics shows a greater intensity to the odd harmonics, and Figure 6.2c makes it clear that the harmonics all decay at different rates. So, in addition to needing a lot of oscillators, we are indeed going to need EGs on *every* oscillator. And creating a harmonically rich sound may require *hundreds* of sine waves. It is becoming clear why additive did not come along until digital instruments and computers were capable of processing large amounts of data in real time.

Organs aside, additive synthesizers are nearly always digital. It is technically possible to do with an analog instrument, but imagine an analog instrument that has hundreds of sine-wave oscillators, each with its own EG, and VCAs to pull it all together . . . physically enormous, expensive, power hungry . . . not terribly practical (although Hammond did build such an instrument called the Novachord, which is detailed in Chapter 1).

Let's start with something a little simpler than a piano note. In Chapter 3 we looked at the harmonic structure of the standard waveforms: triangle, square, and sawtooth. To create a sawtooth using Additive Synthesis, let us look again at the its harmonic structure in Figure 6.3.

Although time consuming, it is possible to construct a customized, digital additive instrument using software like Tassman or Reaktor, but the rewards may not justify the effort, especially because you may need hundreds of oscillators to even come close to emulating this piano note. In fairness, the sound designer's goal is more often than not creating a *unique* timbre, one that does not attempt to sound like an acoustic sound, and that is where Additive Synthesis becomes arguably more useful and interesting . . . and fun.

Instruments

The level of control and large number of oscillators needed relegated Additive Synthesis for many years to the studios of academia, where sophisticated, expensive computer systems were available. That level of control would become a little more accessible to the individual user but still needed to be coded using complex applications like Supercollider, Csound, and MaxMSP that take a programming-like approach. A user-friendly interface that would appeal to someone uncomfortable with coding was still off in the future.

In 1986, the first taste of Additive Synthesis in a dedicated hardware package emerged with Kurzweil's K150. Through what Kurzweil dubbed "Fourier Synthesis," users with an

Creating Sounds from Scratch

Figure 6.4
Kawai K5000 additive synthesizer (photo courtesy of Kawai America Corporation).

Apple IIe (pre-Mac) computer connected to the 150 allowed its 240(!) oscillators to be shaped with envelopes and tuning control.

A decade later, in 1996, Kawai released a more refined workstation called the K5000 that had fewer additive oscillators (64) but had hundreds of samples that could be layered, a more sophisticated filter section, LFOs, an arpeggiator, *and* a forty-track sequencer (see Figure 6.4).

Resynthesis in Additive Synthesis

Building a sound from scratch in Additive Synthesis can be labor intensive. When emulating an existing sound it can be useful to use spectral analysis tools (as we used earlier in the chapter) that may already be present in your DAW. At the very least it can be educational: How better to understand how a motor works than to take it apart and put it back together—hopefully without any leftover parts! A more practical and user-friendly approach to this deconstruction involves a process we will see again in a wavetable called *resynthesis*.

Resynthesis exploits the computer's efficiency at quickly analyzing a complex sound sample and generating the data needed for its reproduction and manipulation in an Additive Synthesis environment. Once analyzed, a facsimile of the imported sound is organized into its essential harmonics—now produced by dedicated sine-wave oscillators,

Figure 6.5
Screenshot of Logic's Alchemy.

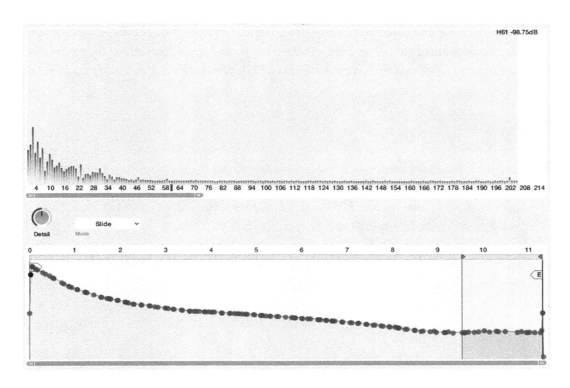

Figure 6.6
Screenshots of Alchemy showing (a) pan location and (b) phase of each harmonic. As mentioned earlier, the sample we resynthesized is in stereo, so there are harmonics that are more prominent on the left-hand or right-hand side of the stereo panorama, and that information, too, is captured and available to be customized.

rather than simply replaying a sample—with each harmonic *independently* following a volume envelope shaped according to details from the analysis. Stereo captures may also include information about where in the left–right panorama a given harmonic was located, and its phase at the start of the sound. A great instrument for studying this process is Logic's Alchemy (originally created by Camel Audio) (see Figure 6.5).

Further shaping to create more depth and stereophonic contrast can be done with tuning of individual harmonics, shifting its pan location, and the phase of each harmonic (see Figure 6.6). The level of control is enormous.

Manipulating Resynthesized Sounds

Once a sample has been resynthesized, the fun begins. Figure 6.7 uses the example of a resynthesized acoustic guitar plucked string. The harmonic structure has been left intact but the volume envelope of the lowest three harmonics has been shaped with a dip, giving the attack a double-hit quality; the first several harmonics have been panned more

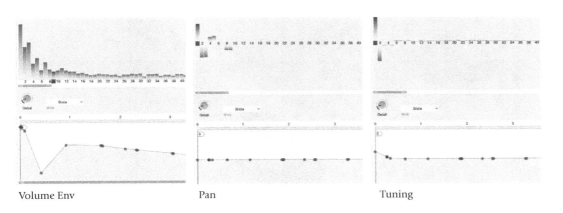

Figure 6.7
Manipulation of a resynthesized acoustic guitar plucked string in Alchemy.

dramatically left–right than what was in the original sample, resulting in a wider-than-natural guitar sound, and the tuning has been shaped with its own envelope so the first harmonic starts a little sharp, and the second a little flat, and then quickly is restored to the correct pitch.

Spectral Synthesis

Whereas Additive Synthesis is very much a calculated, methodical approach to constructing a timbre, spectral synthesis is far more abstract. Assembling a sound in spectral synthesis using an instrument like Alchemy initially resembles drawing a bit more than it does music synthesis. Take a look at the example in Figure 6.8.

The timbre in Figure 6.8 is one that has two distinct yet compatible sounds. It is a stereo patch with the top half of the image representing the left channel, the bottom the right channel—any variation between the two will create the necessary left–right contrast needed for stereo separation. Most of the image is black, representing no sound. The streaks and blocks are where sound will be created at intensities relative to the brightness of the shade.

In a process that really is a lot more like drawing with a computer mouse, the user selects a tool (see palette in Figure 6.9) from a pop-up list that will shape the desired tone/effect. It might be the *circle* or *blurred circle* (which are seen in Figure 6.9) that create

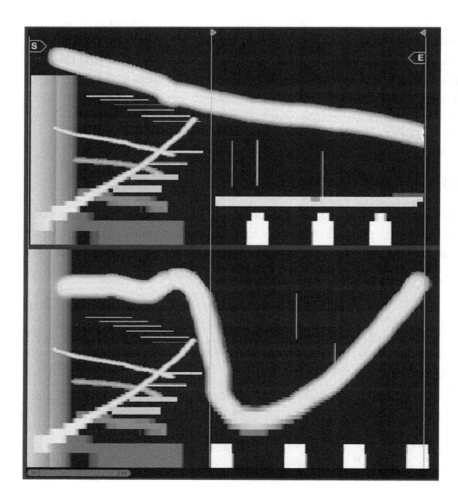

Figure 6.8
Screenshot of spectral synthesis in Alchemy. The x axis represents time, the y axis is frequency, and the intensity is represented from a grayscale range of black (off, no sound) through increasingly brighter shades of gray, to white (highest amplitude).

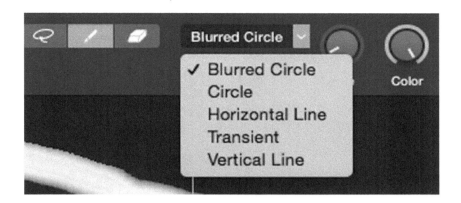

Figure 6.9
The tool palette in Alchemy's spectral editor. Along with the drawing tools, to the left you can see a Lasso tool used for soloing a manually selected region of the spectrum (in Mask mode), and an Eraser tool for removing something that is, well, not working out.

spray-paint-like streaks across the spectrum. There are horizontal lines that are helpful for holding a steady pitch, vertical lines serve as noise bursts, and transients that stretch across a wider range of the frequency spectrum.

Atmospheric sounds are the easiest to create with this technique, but it is possible—especially using the Horizontal Line tool—to build up harmonics in a more conventional additive fashion. Alchemy also allows layering of four timbres, so an atmospheric effect made using spectral synthesis can easily be layered with a sample or wavetable or subtractive patch . . . or all three!

Iris: Is it Sampling? Subtractive Synthesis? Spectral Synthesis? Yes

The last technique before we dive into the Producer's Point of View is an instrument that is hard to classify. In some respects it is like the spectral synthesis of Alchemy but it also has elements of subtractive synthesis and sampling. Without a clear category home, we drop it in here. The instrument in question is Izotope's Iris.

Izotope has made a name for itself authoring signal processing and noise-reduction tools that have become essentials in many recording studios. Their noise-reduction software "RX" displays a spectrogram on which engineers can see a visual anomaly that is likely to be an offending noise they are trying to eliminate (see Figure 6.10). With user-friendly tools, that region can be selected and removed with no or little trace of the offending blemish. Could this be used to make an instrument? Absolutely.

So, taking an approach that was originally intended for restoration processes, sound designers can use the tool creatively to select only the parts of the more complex sound that is desired. Sounds like subtractive synthesis. Yes, but the approach is very much additive as well, especially if your starting point is a sample that covers a wide portion of the audible spectrum. The process is a lot like placing a large piece of paper against a gravestone and rubbing it with a pencil. The etched portions of the stone will be omitted because there is less friction, so the pencil's graphite doesn't leave much of a trace. Iris uses the sample—either one provided or one that you upload—as the granite monument, and you use the available tools to reveal the portions you want to hear.

Creating Sounds from Scratch

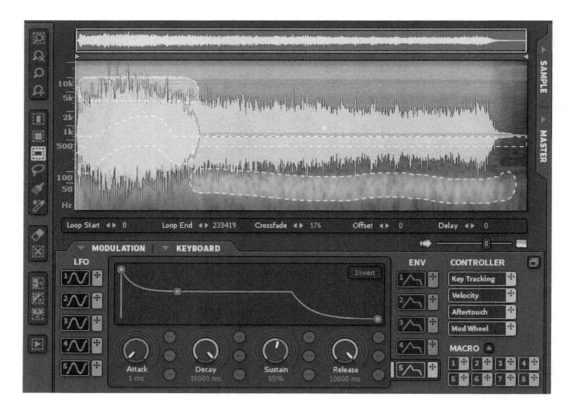

Figure 6.10
Izotope's Iris 2 with a "Spraycan" sound sample filtered to create an atmospheric effect. The dotted lines outline the segments of the original sample that will be audible—every other part of the sample is muted.

The Producer Point of View on Additive Synthesis

Although Additive Synthesis started as a way to recreate existing sounds thorough careful addition of sine waves set at different harmonics and amplitudes, nowadays it has evolved into a complex and sophisticated way of creating incredible original sounds and patches. Among the most used additive hardware synthesizers for music production we must mention the New England Digital Synclavier, the Digital Keyboards Synergy, the Kurzweil K150, the Kawai K5, and the Kawai K5000. All these machines were very powerful for the time they were introduced to the market, moving the boundaries of electronic production a step further every time. For a quick comparison among these important synthesizers and their key features look at Table 6.1

After the K5000, most of the hardware synthesizers moved to either sampling and pulse-code modulation (PCM) or to hybrid synthesis techniques that would combine several different approaches to sound creation. Additive Synthesis has made a big comeback though with software synthesizers. This is due mainly to the fact that this type of synthesis requires a powerful CPU for generating in real time the necessary sine waves and their envelopes in order to create complex and interesting sonorities. Software synthesizers offer the ideal solution to this problem, because they run on very powerful computers, capable of handling several hundreds of waves simultaneously. In addition, the fact that we can take advantage of large screens to create, add, and edit the harmonics that constitute the core of an additive patch, helped the evolution of incredibly complex additive software synths, capable of creating some very exciting new sonorities. There are several software synthesizers these days that feature Additive Synthesis, in most

Table 6.1 Overview of the Most Important Hardware Additive Synthesizers

Hardware Synthesizer	Year of Production	Key Features	Comments
Synclavier	1977–1984	• Built-in sampling, FM and Additive Synthesis– Later models had music notation printing, sequencing, and hard-disk recording capabilities • One of the first workstations	This was the top of synthesis starting from the late 1970s up to the mid-1980s. Artists like Sting, Genesis, Frank Zappa, The Cure, and Michael Jackson used this on their albums.
Digital keyboards Synergy	1982–1985	• Combination of additive and phase modulation synthesis • Built-in four-track sequencer • Expansion cartridges	This was a very versatile synthesizer with patches that ranged from electric pianos, to synth woodwinds, to strings, organs, and organ-like pads. It was used extensively by Wendy Carlos on albums such as "Digital Moonscapes" and "Beauty and the Beast".
Kurzweil K150	1986	• Very powerful • Splitting keyboard • Extremely editable • Full MIDI implementation	For the period, this was an excellent machine. Its piano sounds seem to be some of the favorite patches among users.
Kawai K5	1987	• Very digital in nature but with an "analog" approach to filters • Hard to program • The K5m was a rack version of the K5	This was a very powerful machine but hard to program. Some of its most successful sounds were pianos, brass, electric pianos, organs, synth leads, and synth basses.
Kawai K5000	1996	• Combines additive and PCM synthesis • Extremely powerful and fairly easy to program • Built-in sequencer • Also available in rack version	This machine was one of the last attempts to market a hardware-based additive synthesizer. It was a great machine but never very successful commercially. Among some of its signature patches are complex and evolving pads and leads.

Figure 6.11
Example of additive software synthesizer: Alchemy's (Logic Pro X) three main sections.

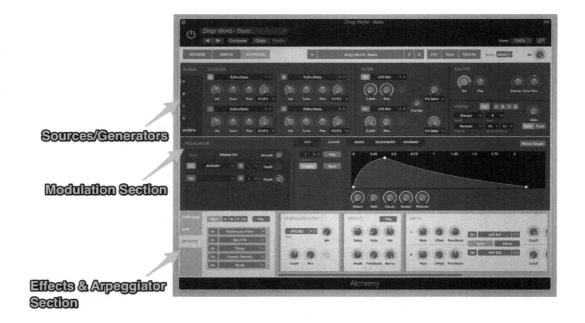

cases along with other types of synthesis such as FM, Sampling, or Granular. Some of our favorites are Alchemy (now bundled with Logic Pro X) (Figure 6.11), Loom by Air Music Technology, and Morphine by Image-Line.

The majority of the additive software synthesizers allow you to easily control not only the amplitude of each harmonic, but also tuning, panning, phase, and envelope. As is always the case for software applications, we recommend keeping an eye on new titles and updates. Things move very fast in this area! The landscape of soft synths that either feature or include Additive Synthesis is pretty open, and there is not a single one that would most likely fit every producer's needs. Each one seems to appeal to different areas or styles. Therefore, when looking for good additive software synthesizers we recommend looking at some of the specifications and features and decide if it is the right for you. Table 6.2 is a list of some of the key features that we recommend considering when looking for a new additive synthesizer (see also Figures 6.12–6.14).

As we mentioned, the options for software synthesizers are many and keep chaining rapidly. Nevertheless, we list some of our favorite ones and their key features in Table 6.3.

Sonic Categories and Additive Patches

As we learned in the first part of this chapter, theoretically with Additive Synthesis any complex waveform can be dissected in a series of different sine waves at different frequencies and amplitudes. This approach has allowed synthesizer manufacturers in the past to create some fairly realistic acoustic sounds (such as the piano of the Kurzweil K150). The reality is that, nowadays, for the modern producer, Additive Synthesis can require some complicated programming and many hours of trial and error in order to reproduce some convincing acoustic sonorities. In addition, the advances made by both sampling and physical modeling make, in our opinion, Additive Synthesis a much better suited tool for creating original synth pads and leads than quasi-realistic acoustic patches. As we learned in the first part of this chapter, it is fairly easy to recreate geometrically shaped

Table 6.2 Key Features of a Software-Based Additive Synthesizer

Features	Comments
Easy-to-use interface	Look for a pleasant and intuitive interface. This type of synthesis is already complex on its own, and you don't need any extra overhead when learning or operating the synthesizer.
High number of partials per voice (256 or higher if possible)	This is crucial for having a powerful and flexible engine
Easy to edit and control harmonic spectrum	At the base of this type of synthesis is the ability to add different harmonics at different frequencies (Figure 6.12). If the interface for doing this is kludgy or not fluid, then the process will be tedious and unpleasant.
Additional type of synthesis engine	Most additive synthesizers offer additional engines such as FM, sampling, and granular. Having these extra options gives you an almost infinite choice for making original sounds.
Additional effects	A good selection of effects such as reverb, delay, filters, amp simulation, and modulation (chorus, flanger, etc.) is essential for creating new and original patches (Figure 6.13).
Flexible modulation section	This is one of the most important features. Being able to access a flexible modulation matrix is crucial to obtaining interesting results. Look for a good number of LFOs assignable to any parameter of the synth engine (Figure 6.14). Also look for modulation sources such as envelopes, sequencers, note parameters (such as velocity, aftertouch, etc), and external controllers (such as MIDI CC).
Resynthesis	Resynthesis is a great tool for creating original variations on preexisting waveforms (acoustic or synthesized instruments). We love having an additive software synthesizer that allows us to apply the infinite flexibility of Additive Synthesis to acoustic waveforms

waveforms such as triangular and square by using Additive Synthesis. Now let's think outside the box a little bit and unleash our creativity with Additive Synthesis!

Pads

With Additive Synthesis we can create some exciting pads with a fairly wide range of sonorities, ranging from smooth and voice-like, to more edgy and sharp ones. To create original and interesting pads with Additive Synthesis the first thing to do is to take control of our sources. Depending on the synthesizer you use you can have one, two, four or more sources. Think as these as different channels of a mixing board on which you have full control over the type of waveform, volume, pan, modulation, and effects. Basically think of the sources as ways of combining, layering, and mixing different sonorities. Although you could use only one, the fact that we can access more than one gives us

Creating Sounds from Scratch

Figure 6.12
Window with amplitude of the harmonics in Morphine by Image-line.

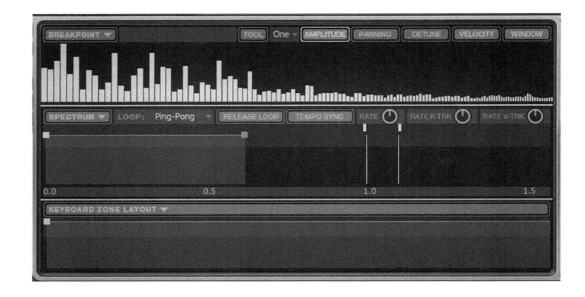

Figure 6.13
Audio effect list in Alchemy.

Figure 6.14
Modulation section in Loom by Air Music Technology.

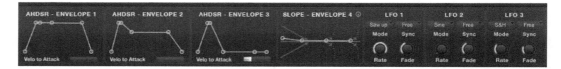

much more flexibility for creating complex and evolving sounds. The typical structure of a multisource Additive Synthesis synthesizer is the one shown in Figure 6.15

Of course, depending on the specific software you use, things might change slightly. The way we recommend approaching sound design for a software-based Additive synthesizer is to start simply first, and then get gradually more complicated by adding more sophisticated

Table 6.3 Some of Our Favorite Additive Software Synthesizers and Their Key Features

Software Synthesizer	Features/Comments
Alchemy: Logic Pro X	• Bundled with Apple's Logic Pro X • Excellent choice of bundled patches • Integrated with Logic Pro X • Clear and elegant user interface • Good balance between power and ease of use • Built-in arpeggiator and effects • Excellent sound-morphing capabilities • Capable of importing EXS-24 patches
Air Music Technology: Loom	• Combined and integrated user interface allows you to see many parameters in a single window • Good amount of bundled patches and presets • Patch randomizer • Sound-morphing capabilities • Up to 512 harmonic partials per patch • Onboard audio effects
Image-Line: Morphine	Resynthesis Great morphing capabilities Up to thirty-two-note polyphony Good selection of bundled patches

elements later. For example, let's work on a pad patch with a voice/organ character. The first step would be to work on the waveform of our first source. We recommend working on one source at a time. This will help us shape each source, depending on our needs and goals. When working on the waveform of a source you have several options, depending

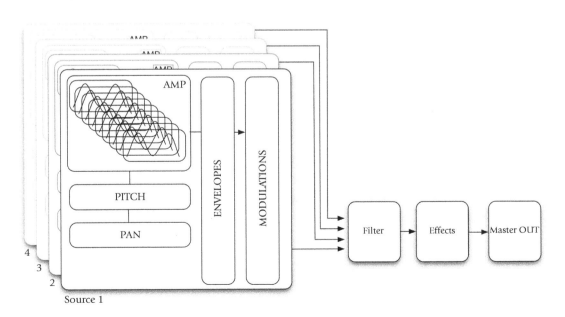

Figure 6.15

A typical structure of an additive software synthesizer.

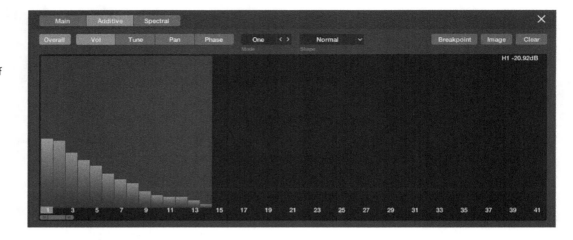

Figure 6.16
A complex waveform generated by adding a series of overtones to the fundamental sine wave in Alchemy.

on the software you are using. You can start from a basic sine wave and then construct more complex waveforms by adding the harmonics at different intervals and amplitudes. This is how a real Additive Synthesis "master" would begin. We like this approach because it doesn't constrain you and it doesn't point you in any specific directions. You are free to experiment and "draw" your own complex waveform by simply adding harmonics. In Figure 6.16, for example, we simply use the mouse to create a simple repetitive waveform that has all the harmonic overtones up to fourteen.

The resulting waveform can ben seen in the oscilloscope screenshot in Figure 6.17. By the way, we highly recommend using always a software oscilloscope when creating any Additive Synthesis patch. It is great tool to have a visualization of where you are heading when working on the harmonics balance of the waveform.

As you can see, this waveform is a hybrid between a sine and a sawtooth wave. You can listen to this waveform in Audio Example 6.1. At this point feel free to experiment with the overtones and their relative amplitudes. In general, for brighter and more aggressive tones add more overtones and higher aptitudes. Just out of curiosity, try to add higher

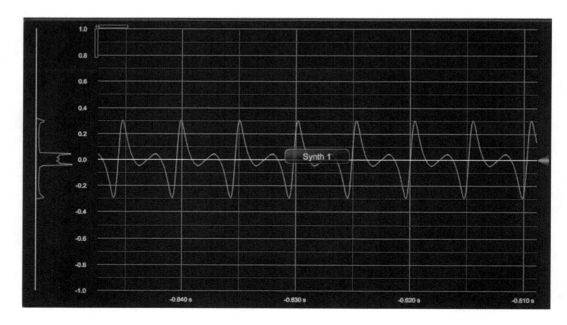

Figure 6.17
The oscilloscope chart of the harmonic structure shown in Figure 6.16.

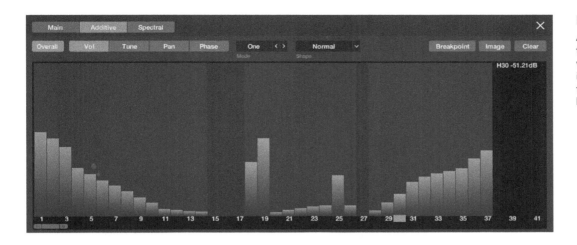

Figure 6.18
A variation of the original waveform shown in Figure 6.16 with the addition of higher overtones.

overtones with a medium-to-high amplitude and listen how this will affect your sound. In Audio Example 6.2 we altered the overtones, as shown in Figure 6.18.

Listen to Audio Example 6.2. Now look also at the oscilloscope graph of this waveform (Figure 6.19). See how more complex it is?

Once you are happy with the waveform of your first source, it is time to work on its other parameters. You should focus on the envelope, pitch, and pan of each harmonic. This is one of the most exciting parts of Additive Synthesis. Knowing that you have individual control over these parameters for each harmonic opens up an almost infinite world of original sounds! Let's take a look at how we can use these parameters to spice up your sounds.

Harmonics' Envelope

In some of the most advanced Additive synthesizers (like Alchemy) you have control over the envelope and loop points of each harmonic forming the waveform of your source. In Figure 6.20, for example, we changed the envelope of the seventh harmonic to have a slower attack and a slower release with a higher peak in amplitude overtime.

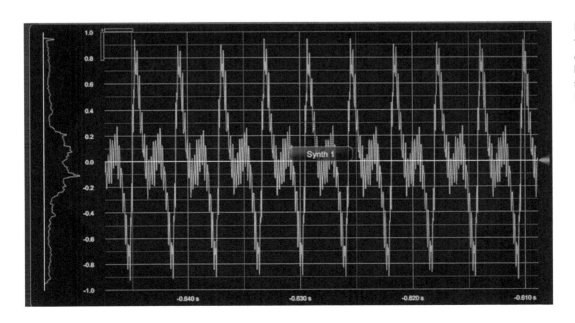

Figure 6.19
The oscilloscope chart of the harmonic structure shown in Figure 6.18.

Figure 6.20
The altered amplifier envelope of the seventh harmonic.

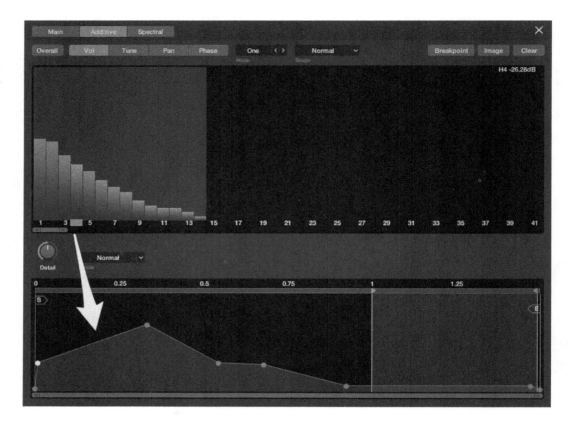

This will result in slightly evolving sounds, with the brightness brought by the slow attack of the seventh harmonic coming in with a slight delay. Listen to Audio Example 6.3 to hear how this envelope change affects the sound. Using the individual envelopes for each harmonic can really give the illusion of an evolving and more complex sonority. Imagine a patch for which each harmonic is added one after the other, creating a tone that becomes more and more complex over time!

Harmonics' Pitch

Another extremely interesting parameter we can work on at the single harmonic level is the pitch. Controlling the individual pitch alteration for each harmonic allows us to move away from the pure character of an Additive Synthesis patch and venture into the more exciting land of nonharmonic overtones. We recommend being pretty conservative with the pitch alteration because too much variation can dangerously mask the pitch of the fundamental. In Audio Example 6.4 we added some pitch deviation for harmonics numbers five, seven, and nine, as shown in Figure 6.21.

Harmonics' Pan

In the same way we are able to change the pitch of each harmonic for our sound source we can also alter the positioning over the stereo image. This is again a great tool for spreading a patch in stereo. The most interesting aspect of this option is the fact that the spread is not based on time (as it would be if we assign the pan to a LFO, for example) or keyboard zone, but instead it is solely based on the harmonic number we pick. Imagine being able to move every other harmonic on a different side of the stereo image! In

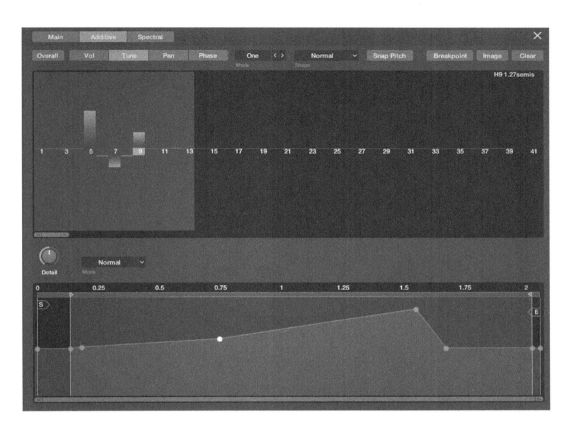

Figure 6.21
The altered pitch for the fifth, seventh, and ninth harmonics.

Figure 6.22 we panned odd and even harmonics alternatively to the left and right of the stereo image.

To better comprehend the possibilities of harmonic panning listen to Audio Example 6.5 to hear the panning shown in Figure 6.22.

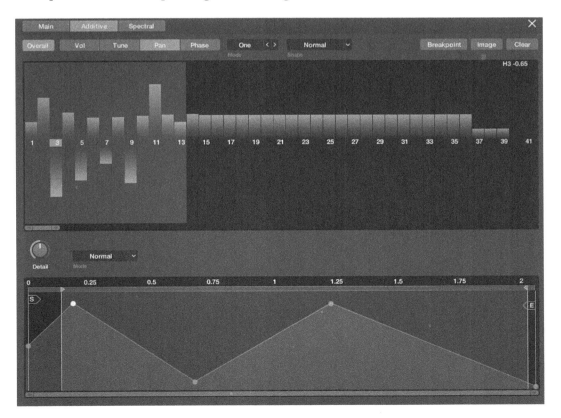

Figure 6.22
Alternating pan settings for odd and even harmonics.

Creating Sounds from Scratch

Adding Other Sources

As we learned earlier, most of the Additive Synthesis synthesizers provide more than just one source. Therefore, what we just did for the first source can be done also for the other additional ones. This option gives us a lot of flexibility when programming a new synth pad. What really makes a pad interesting and captivating is the fact that it can evolve over time, with additional timbres and frequencies delivered after the key note trigger is pressed. Although a similar effect can be also achieved with the envelope of a filter, the capability of having other sources with additional harmonics being generated at different stages of the life of a patch is specific to a multisource additive synthesizer. Because we have control over the individual envelope for each harmonic, we can have a second source featuring a high harmonic that peaks after the main body of harmonics (generated by the first source) is played. For example, in Figure 6.23 we can see the harmonic structure of sources 1 and 3 in Alchemy.

As you can see in Figure 6.23, the second source features only one harmonic in addition to the fundamental. The interesting aspect of the second source though is the fact that the envelope of harmonic 24 has a slow release and peaks much later than all the harmonics of source 1. This creates an interesting effect of a sound that evolves over time. To confirm that this is working, look at the spectrogram in Figure 6.24 while listening to Audio Example 6.6. Notice how the peaks in the high frequencies happen later in time in comparison with those in the main body of the sound.

Of course you can use a similar technique for the other additional sources. In general, we like to think ahead when we work on a new sound and break it down into different sonic components that then we will assign to different sources. For example, we could have source 1 provide the main body of the patch, and sources 2 and 3 could add interesting additions to opposite sides of the spectrum, Source 4 could be then used to bring in elements of surprise by using different envelopes for different new harmonics. Using this approach, we added two more sources to the ones shown previously in Figure 6.23. Specifically, for this pad patch we added two more sources (Figure 6.25). Source 2 covers an area of the spectrum that is higher than the one covered by source 1, with harmonics 32, 33, 38, and 39 peaking a bit later than the rest of the harmonics.

We have used source 4 to "spice up" the recipe. Its harmonic structure is very simple, but we added a bit of overtime pitch alteration to the fifth harmonic. Listen to Audio Example 6.7 to hear the four sources together. As you can hear, our sound is getting more and more complex and interesting.

To the Next Level With the Use of Effects

Now that we have our main sources and our basic waveform set, it is time to start working with some audio effects to further shape our patch and bring it to a higher level of complexity. Most high-quality software synthesizers offer some sort of audio effects palette. Usually some of the most common effects offered are reverb, delay, filters, modulation (chorus, phasing, flanger), and distortion (or amp simulation). We like to consider this step of sound design as a completely new phase of the construction process of a new patch. This phase is comparable to the mixing stage of a recording session. Once we have recorded our tracks (the creation of the waveform[s]), we can go ahead and work on the sound from a "mixing" engineer's point of view. In this phase we have

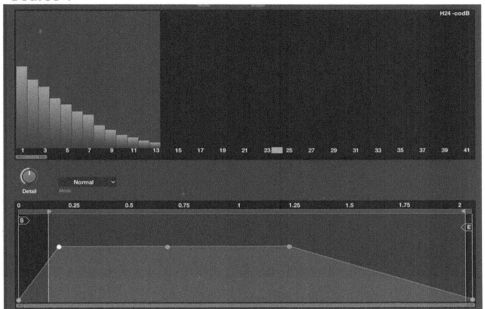

Figure 6.23
Harmonic and envelope structure of two sources in Alchemy.

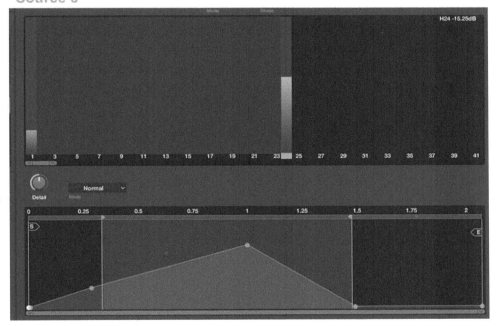

a lot of freedom to experiment and to create sounds that can evolve considerably from the initial raw waveform we were working on. Before we get into the details of specific effects, it is important to understand how each effect category can affect and shape our patch (Table 6.4).

So, how can we use audio effects to further enhance our patches? The preferable approach is to have first very clear in your mind in which direction you would like to bring your sound. For example, maybe you want to make it more direct and with more impact, or maybe you would like to make it more ethereal sounding and distant in space, or maybe you would like to transform it totally and make it more aggressive and original. These are

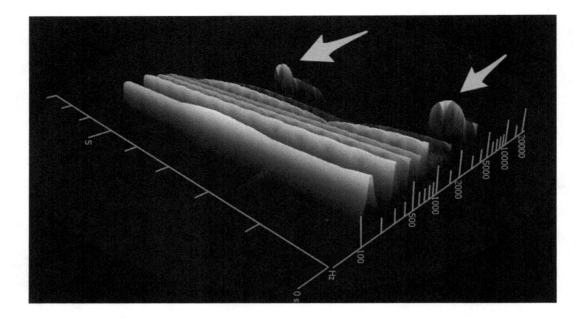

Figure 6.24
Spectrogram of the two sources shown in Figure 6.23.

all possible scenarios (and there are many more) that you should consider before start using the effects of your synthesizer. In order to make this decisional process easier you can use the information listed in Table 6.5 as a starting point.

Now that we have a better idea about how to achieve specific results through the use of audio effects, let's experiment with some of the categories listed in Table 6.5. In the next section we alter the same sounds by using audio effects to create new sonorities that describe some of the sonic options listed in Table 6.5. Let's start!

The Basic Patch

The first thing that we need to experiment with the audio effects section of our synthesizer is a basic patch that we will use for the different examples. For this example we use the patch that we have designed in this chapter and that you can listen to, without any audio effects, in Audio Example 6.8. As you can hear, this is a plain patch that we designed earlier. There are no effects applied at the moment. Try to describe the sound. Is is light, heavy, dry, wet? Is it close to you in space or far away?

High Impact and Direct

Now let's try to make it sound more direct and with a higher impact. Even though this patch (being so dry) is already fairly direct, we can give it more punch by adding the following effects (Figure 6.26):

- Compressor: Mid-compression with the addition of a "Phat" coloring. We used a 0.9-s release time to extend the compression further into the release of the sound.
- EQ: We used a three-band EQ. For the first band we set up 100 Hz, + 6dB, Q 1.6 Oct. For the second band we have 3 kHz, +2 dB, Q 1.2 Oct.

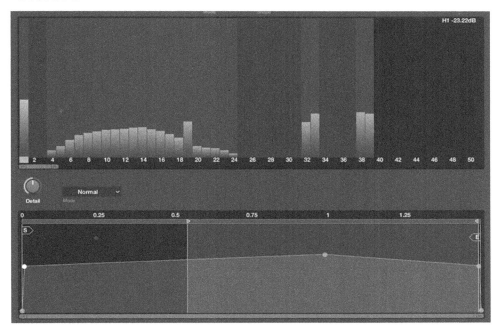
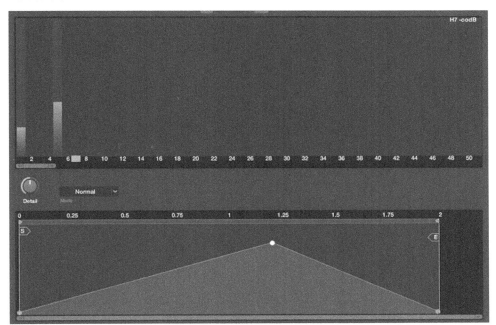

Figure 6.25
Harmonic and envelope structure of the other two sources in Alchemy.

- <u>Distortion</u>: We have a tube amplifier simulator, with a low pregain, and we added a bit of a "nastier" crunch via the "Mech" control in Alchemy.

Listen to Audio Example 6.9 to hear the impact that these effects have on the original sound.

Table 6.4 Main Characteristics of Sound Effects

Type of Effect	Impact on the Sound	Comments
Reverb	It allows you to place the sound into a specific space. The space can be real (room, concert hall, church etc.) or totally artificial.	Reverb is an integral part of sound design and sound creation. In general, adding reverbs allows us to move the sound farther away from the listener, resulting in a less direct and "in-your-face" effect.
Delay	It can create a sense of distance from the sound source. It can also be used in a very creative way with delays that function more as sound design elements than realistic effects.	Delay is one of our favorite effects to create more interesting and surprising sounds. In particular, you can use delay to open up a sound on the stereo image by having alternating bounces on different sides of the stereo image.
Filters	Very effective and powerful tools to carve and shape the frequencies of the basic waveform.	As we learned in the subtractive synthesis chapter, filters are the key tools for further shaping a waveform. Because filters can have their own envelopes, we can use them to change the sound over time very effectively.
Modulation	Modulation effects, such as chorus and flanger, have the basic characteristic of "liquefying" the original sounds.	These are excellent tools to add more substance to a weak and thin waveform. They can add character and at the same time they can also smooth off some edginess.
Compression/ expansion	Compression can add punch and directness to our waveform. Expansion can be used to add more dramatic dynamic changes.	If used correctly compression can really add punch and presence to a sound.

Table 6.4 Continued

Type of Effect	Impact on the Sound	Comments
Limiter	Used to increase the overall volume of a sound without adding distortion.	It is basically a limiter with a ratio set to 1:100 or higher. In sound design it can be useful to increase the volume of a waveform without distortion
Gate	The noise gate allows us to set the amplitude of a waveform to a set level when the signal goes below the set threshold. If the level is set to negative infinity then we will have complete silence when the amplitude drops below the threshold.	It can be useful in specific cases for which you want to have the sound go on and off, depending on its amplitude.
Distortion	It adds a level of simulated analog or digital distortion to the signal. Overall it can be used to add character and punchiness to a waveform.	Distortion can be embedded in "amp simulators," which also adds extra coloration to the original sonority.

Liquid

To morph our original patch into something more fluid and liquid, we start adding a modulation effect, followed by a three-band equalizer. To add a bit more of a wet touch, we also add a long reverb at the end of the chain. Here's a detailed list of the effects for this patch (Figure 6.27).

- Modulation: We use a light flanger with a small depth and fast rate to avoid being too present. We use also a little bit of feedback (around 50%) to make the modulation effect richer.
- EQ: We use two bands: one set at a frequency of around 250 Hz, with −10dB and a Q point of 1.6 Oct, and the second band set to 5 kHz, with a gain of 6–8 dB and a Q point of 1.6 Oct.
- Reverb: We use a long reverb (5–6 s) with a large-size room and a medium diffusion.

Listen to Audio Example 6.10 to hear how these effects can add a sense of fluidity to the original patch.

Distant and Ethereal

The impression of a distant and ethereal sound can be achieved with a long reverb with a dry:wet ratio higher than 70% in order to hear more the wet sounds than the dry ones. A long delay will also help to position the sound in a larger and more distant space. You

Table 6.5 Use of Effects to Achieve Specific Sonic Results

Sonic Results	Effects to Use	Comments
High impact and direct	• Compressor • Light saturation/distortion • EQ: boost low frequencies between 60 and 180 Hz • For more presence boost frequencies between 2 and 8 kHz	Try to avoid long reverbs or delays if you are looking for more impact. The compressor can really help if used correctly and with overdoing it. Too much compression can take the "life" out of your sound.
Liquid	• Modulation effect such as chorus or flanger • Cut some low frequencies below 200 Hz • Boost gently frequencies between 8 and 10 kHz • Reverb: use a little bit of a long reverb without overdoing it	Try to add the chorus after the reverb in order to "liquefy" also the reverb tail
Distant and ethereal	• Reverb with long decay time (5 s and up) • Long delay • Gentle rolloff of frequencies below 300 Hz	To create the effect of something coming from far away and floating in midair, long reverbs and long delays are great tools. If you are looking to give the impression of a sound that is not connected to the ground, roll off some low frequencies with a HPF. The higher the cutoff frequency of the filter the less "connection" with the ground the sound will have. To enhance the impression of a wider space we like sometimes to add a second reverb (generated by a different plug-in) at the end of the effects chain.
Sharp and edgy	• Low to medium distortion • EQ boosting frequency between 4 and 8 kHz • Lower the bit resolution to 14 or 12 bits	Using a low-fidelity digital filter in conjunction with a little bit of distortion can add a nice edgy characteristic to the sound. For extra sharpness use a peak filter with a medium Q point and a 2- to 4-dB gain in the range of 4 to 8 kHz

Table 6.5 Continued

Sonic Results	Effects to Use	Comments
Disruptive and aggressive	• Heavy distortion • Addition of an amp simulator • Boost the mid–high frequencies using a peak EQ between 4 and 8 kHz • Lower the bit resolution to between 12 and 8 bits	The goal here is to add some "crunchiness" to the sound. Distortion (analog and digital) can really change the characteristic of the patch. Try to use the EQ aggressively by using a gain between 5 and 7dB.
Boomy and heavy	• Boost frequencies between 60 and 100 Hz • Add a sub-bass plug-in to generate additional low frequencies • Compressor with mid settings	This category applies mainly to basses, bass drums, and low pads. Adding a subfrequency generator really helps to add that extra bottom to your sound.
Thin and icy	• HPF with a cutoff frequency between 500 Hz and 2 kHz • A touch of reverb with a long decay time (3 s or higher) • HPF on the reverb tail with a cutoff frequency between 2 and 4 kHz • A touch of flanger	Cutting the low end of the spectrum helps lift the sound and at the same time gives a sense of coldness. The light reverb with the HPF also contributes to detaching the sound from the ground and at the same time making it "frozen" in space.

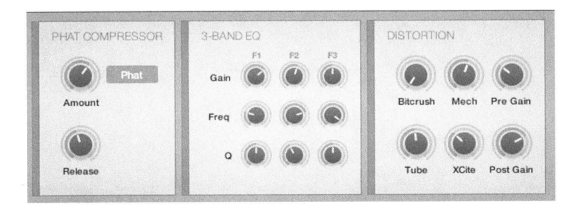

Figure 6.26
Effects for a high impact and direct patch.

Figure 6.27
Effects for a liquid patch.

Figure 6.28
Effects for a distant and ethereal patch.

can use a HPF with a cutoff frequency of anywhere between 100 and 300 Hz to thin the sound a little bit. For this example, we also added a second convolution reverb at the end of the chain in order to add a wider space to the sound (Figures 6.28 and 6.29). These are the effects we used for this patch.

Figure 6.29
The additional reverb added at the end of the effects chain for distant and ethereal patch.

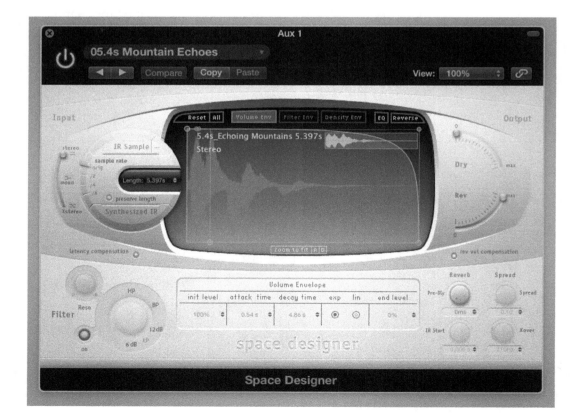

- Synthesized reverb: We used a long reverb with a HPF set around 500 Hz, with no high-frequency dumping and a ratio of wet:dry of around 80%.
- Stereo delay: A stereo delay helps widen the stereo image and at the same time move the sound into a larger space. For Audio Example 6.11 we used different rates between the left-hand (two-beat rate) and the right-hand (one-beat rate) sides. A HPF with a cutoff frequency set at 650 Hz to help shin out the delay bounces.
- HPF: A HPF set at 450 Hz filters out unwanted frequencies that would make the patch heavier than needed.
- Convolution reverb: We used the Logic Pro X Space Designer to add a more natural and wider space. For Audio Example 6.11 we used an outdoor impulse response with a reverb length of 5.4 s (Figure 6.29).

Listen to Audio Example 6.11 to hear how these effects can make your original patch more ethereal sounding.

Disruptive and Aggressive

As a final example of how to use audio effects to further enhance your original sounds, we would like to take a different direction and go more toward an aggressive sonority. For these types of patches, we prefer working with a sound that has a sharper attack and shorter release. So we altered our original patch by simply reworking the envelope of the amp by shortening the attack and the release. Listen to the new sound without effect in Audio Example 6.12. Notice how the addition of an arpeggiator gives a nice rhythmic character to the patch.

To add more bite and aggressiveness to the waveform we added first an overdrive pedal and then a distortion (Figure 6.30). These two pedal effect simulators are great for transforming the original sound into something sharper and more abrasive. To smooth out some of the edginess and allow for a more fluid transition between notes we added

Figure 6.30
The effect pedal simulator in Logic Pro X used for the disruptive and aggressive patch.

a flanger. Finally, to add a bigger space we used a delay instead of a reverb. The delay not only enhances the rhythmic pattern of the arpeggiator, but also preserves the global aggressiveness of the sound.

Listen to Audio Example 6.13 to hear the use of audio effects in action when creating a more aggressive sonority.

As you can see, the use of audio effects is an extremely valuable weapon for creating new original sonorities. Keep in mind that these techniques can, and should, be applied to any type of synthesis. In fact, the simpler the synthesis type, the more the use of audio effects can enhance and inspire your creativity!

Exercises

6.1 Create two original pad sounds using an additive synthesizer. Call them "Additive Pad 1" and "Additive Pad 2."

6.2 Using the harmonic envelopes, pitch, and pan, create a variation of each of the two previous patches. Try to stretch your imagination! Name these two new patches "Additive Pad 1 Variations 1 and 2" and "Additive Pad 2 Variations 1 and 2."

6.3 Now take one of the two patches created in step 2 and create different variations using audio effects that reflect the following four different sonic types: (a) liquid, (b) disruptive and aggressive, (c) thin and icy, and (d) distant and ethereal.

6.4 Now imagine two sonic categories of your own. Alter another patch created in Exercise 6.2 to reflect their characteristics.

7

Sample-Based Synthesis

Introduction

Once hardware synthesizers and eventually software synthesizers were able to offer large memory capacities and processing power sufficient to reproduce multiple recorded sounds simultaneously, sample playback seemed to be the Holy Grail many sound designers and composers had been looking for: an electronic instrument that would convincingly mimic the sound of an acoustic instrument. For some sounds—particularly percussion and instruments with little or no sustain and quick decay—the results were quite impressive, far beyond anything that was previously possible through other means of synthesis.

For all other instruments, however, like winds and bowed strings, the results were less convincing. The sample being replayed was a static reproduction and challenging to make sound musical. Those involved with developing and playing the first sampled instruments quickly gained an appreciation for the complexity of sound generated by acoustic instruments and the nuances introduced by their skilled performers. Not only did a real performance incorporate a change in timbre with every semitone shift up or down, but changes in dynamics, attack, and a seemingly infinite range of performance techniques limited the sampler's ability to achieve an illusion of reality. Consider in your mind's ear the sound of striking a piano key harder and softer and how the harmonic content of the sound changes with dynamics. Then imagine playing the highest C and the lowest A on a piano (the standard top and bottom of the instrument)—the differences in timbre across such a wide range on the same instrument is significant. The shift of a semitone up or down and the intensity with which a key is struck may produce what seems like minute timbral variation, but in the context of a sample set played back across an eighty-eight-key keyboard it will be immediately detectable.

As digital technology evolved with greater memory capacities, faster processors, and larger hard-disk storage (all for less money), it became possible to develop instruments that could store larger numbers of samples at longer durations and at higher sampling and bit rates. Early sampled pianos, for example, were limited to having a single sample span a minor third or more and used for all dynamic ranges. It is not uncommon for modern sampled pianos to have multiple samples *per* semitone representing the timbral differences across the dynamic range, and alternate samples for each key to provide variety when notes are repeated. In general, the greater number of samples, the more convincing the aural illusion (see Figure 7.1).

Figure 7.1
Logic Pro EXS24 Sampler's Edit Map. This screenshot of Logic Pro's EXS24 sampler shows the large number of samples used in making this very convincing emulation of a real grand piano. Each note of the keyboard, from B1 to C7 (seven octaves!), has its own set of *eight* individual samples representing different velocity ranges—that's 688 samples for one 88-key piano!.

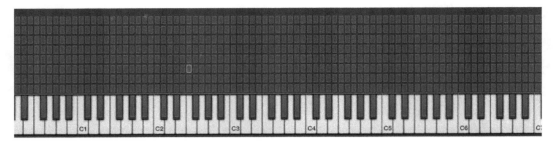

Although advances in digital audio technology made sample playback commonplace in professional and home studios, the first example of an electronic instrument that was capable of reproducing a recorded sample on demand was, arguably, the Mellotron (see Figure 7.2). This instrument was cleverly designed with reels of tape at the ready to play back prerecorded "samples" when a key was pressed. Among the most memorable uses of this instrument is the opening calliope-like segment of "Strawberry Fields Forever" by the Beatles, but there are countless examples in the 1960s and 1970s pop/rock repertoire, and excellent digital samples of the Mellotron are now reintroducing the sound to a new generation. The Mellotron came preloaded with sounds and beat loops, perhaps foreshadowing an approach to music production that would later become commonplace.

Figure 7.2
A modern Mellotron, the M4000 (photo courtesy of Streetly Electronics).

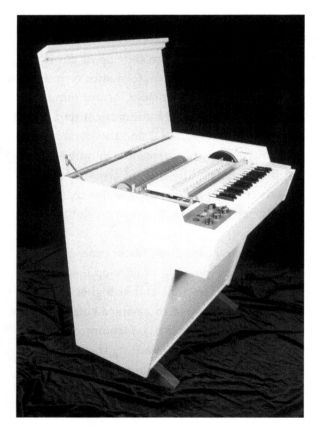

Fortunately, as is the case with so many great instruments of the past, very high-quality emulations of the Mellotron that mimic both the sounds and physical limitation of the instrument can be installed in your DAW of choice as a plug-in. And for those who play live, it's far easier to move a laptop-based Mellotron than its 120+ pound predecessor (unless you have roadies, of course). In addition, software samplers like Kontakt come with excellent Mellotron emulations (see Figure 7.3).

Digital sampling became a practical reality—at least to those with a healthy bank account—when New England Digital introduced the Synclavier around 1980. At the time, of course, digital processing, storage sizes (i.e., hard drives), and memory capacity (RAM) were minuscule

Figure 7.3
Mellotron software emulations: M-Tron Pro (left) and Native Instruments' Kontakt (right).

compared with those of modern standards. For years, professional samplers would be dedicated "turnkey" systems—Macs and Windows PCs, with their graphical interfaces, demanded much of the computer's resources, and thus not yet suitable for the task. In the 1990s, however, that began to change.

Principles of Sample-Based Synthesis

Building your own sample-based instrument using software like Native Instruments' (NI) Kontakt, Apple Logic's EXS24, Avid's Structure, or MOTU's MachFive is relatively simple once you get your brain around the steps specific to the software. The visual interface offered by these applications and many others is far more intuitive than what was available with hardware samplers of the past. Initially, you may be met with a screen that is more of a basic player interface with the editing capabilities buried a little deeper.

The EXS24 interface contains parameters that give it a striking resemblance to that of the other synthesis methods we have explored so far insomuch as we have LFOs, envelopes, and filters available (see Figure 7.4). What is new, however, is that our raw sound is not a standard waveform (such as a triangle, square, or sawtooth) but a recorded sound. Sophisticated sampling instruments offer the usual manipulation tools previously listed as well as real-time control of these characteristics through your MIDI controller—for example by using the modulation wheel, velocity control, aftertouch, and foot controllers to affect LFO intensity, cutoff frequency, etc.

Techniques for Custom Sample Sets

Edit Mapping

Using your own samples to build a sampling instrument involves a process known as *edit mapping*. Clicking on the "edit" button on the top-right of the EXS24 interface will open up the instrument editor where the edit mapping is done. At its most basic, edit mapping is the process of attaching a raw sound that you have recorded or acquired and assigning it to a trigger note. As we see in Figure 7.5, selecting a key (in this case C3 or "middle C") actuates a sample from the list called "Empty Ring - f" to play.

Figure 7.4
The sample-playback interface of Logic's EXS24.

Figure 7.5
Velocity zone editing in EXS24—samples are from a set of sounds recorded off a soda can.

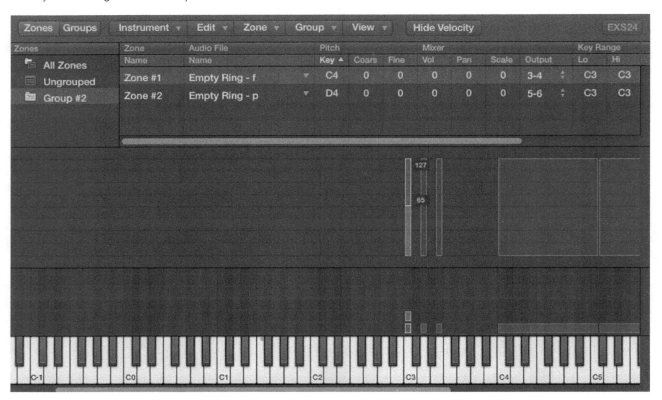

The procedure in EXS24 is similar to what you will find with other software: You can either drag-and-drop a soundfile from the desktop onto the trigger key or manually define a "zone" and import which a sample can be loaded. Notice that the C3 key of the keyboard is darker than the rest, and right above it are two boxes representing velocity zones. These visually reflect the signal being received from the MIDI controller. In this case, the C3 is being held down: The small box above it shows the sample being triggered (also highlighted in the preceding list of samples) and the longer box shows the velocity range within which the key was struck and thus the sample that will be played.

Notice that there are samples loaded onto D3 and E3 labeled as "Fizz Only" and "Ice Roll Texture." If you look closely, you will also see that in addition to "Empty Ring - f" there is an "Empty Ring - p." These samples were recorded with a light flick (p) and a heavier flick (f) of a soda can with an index fingernail. Because they are meant to represent the timbral differences of a softer or harder strike, it is necessary to upload both samples onto the same note and define the range in which each sample will be triggered. In this case, note that the key was struck hard enough so that the velocity was in the 65–127 range; thus "Empty Ring - f" was triggered. Conversely, striking the key with a velocity of less than 65 will result in "Empty Ring - p" being triggered. These definitions are entirely up to you and can be easily adjusted under the "Vel. Range" column in the preceding list. Notice how only these ring samples have Vel. Range checked as On, and you can clearly see in the adjacent columns the limits as previously indicated.

Pitch

When a sound is assigned to a particular note, by default its normal playing pitch will be maintained only at that location. Triggering the sample higher in its allocated zone will play it faster and thus at a higher pitch. This is the default behavior, and the assigned note is referred to as the *root key*. However, if you are importing, say, a percussive sound that you want to be able to move around or spread across several keys independently of pitch transposition, you will need to manually disable Pitch Mode so the root key will be ignored.

With some sounds it may be appropriate to spread the sample across several notes and have it transposed accordingly. Figure 7.6 shows the "Empty Can Ring" sample placed at C1 and stretched up to G2. As long as "Pitch" is selected under the Playback column, the pitch will change as you play different notes within that octave. As mentioned already, the pitch shift is achieved by changing the length of the sample, so if the sample contains a rhythmic pattern or the duration is otherwise important, you will need to build separate samples, each processed with pitch-shifting software (usually built into your DAW) to maintain its proper length.

Audio Example 7.1 is a short composition that uses only sounds from a soda can and a glass with ice.

Groups

Acoustic instruments have a variety of performance techniques—for example, orchestral strings (violin, viola, cello, and double bass) can be either bowed (*arco*) or plucked (*pizzicato*

Creating Sounds from Scratch

Figure 7.6
EXS24 sample spanned across more than an octave so that pitch shifting will be introduced when played anywhere other than the original root note.

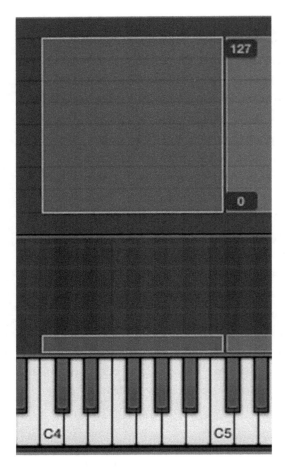

or pizz), they can be bowed near the bridge (*sul ponticello*) or over the fingerboard (*sul tasto*), the wood side of the bow can be tapped or pulled across the string (*col legno*), and so forth. These variations can be stored as separate samples, but it would be far more convenient to jump to them while playing. This is a scenario for which the *groups* function is especially useful.

Let's use the example of a set of double-bass switching between *arco* and *pizzicato* with samples loaded into NI Kontakt.

The mapping editors in Figure 7.7 represent two banks of samples stored within the same double-bass instrument: *arco* samples on the left and *pizzicato* samples on the right. A convenient method for switching between the groups during a performance is to assign unused keys on the MIDI keyboard as *key switches*. The high C through F keys are key switches for seven different sample sets: Sustain (C), *Fortepiano* (C#), *Sforzando* (D), Staccato (D#), Tremolo (E), *Pizzicato* (F), and Staccato 2 (F#). The performance technique required here is playing the bass line with the left hand and pressing the desired key switch with the right to jump to another sample set when needed. A line could have long bowed notes, then a few *pizzicatos*, and back to the bow, all without leaving the same instrument. Obviously this is really effective for live performances but can save a lot of time for the composer/arranger recording individual instruments in the studio.

Figure 7.7
NI's Kontakt Edit Map shown with Arco (left) and Pizzicato (right) sample sets, selectable by the Groups function.

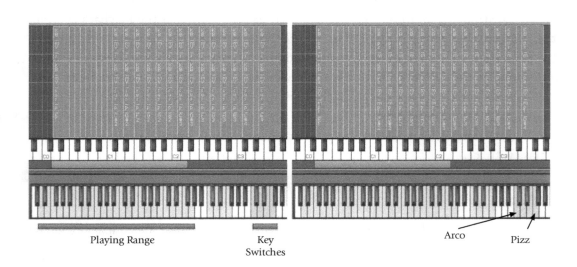

Another use for groups is for using multiple samples of the same note. Why would anyone want to do that? A weakness that exposes an otherwise good-quality sample as coming from a sampler is when two or more repeated notes sound *exactly* the same—something that, fortunately, is not possible on any acoustic instrument, no matter the skill level of the performer. A way to avoid the redundancy is to capture two or more samples of *each* note at *each* velocity range (yes, you may need a bigger hard drive) and programming your sampler to select the sample at random or in a round-robin fashion. This is quite simple to do in Kontakt; A pop-up menu for selecting the behavior of groups includes the options of key switching, random, round-robin, and others.

These two examples of applications for groups are conventional in the sense that they are imitating an acoustic instrument performance. The possibilities are endless, however, when you consider ways of using groups as a creative tool—for example, assigning multiple samples of contrasting sounds to a single note and having the sampler choose at random what to play. Or inserting key switches *within* the set of keys where the samples are located so that hitting a certain note during performance changes the samples being played, yet another example in synthesis in which you are limited only by your creativity.

If the idea of creating all of the samples on your own is daunting, fear not; there is a glut of excellent sample sets on the market, and customization for your needs will be well within your ability after you have worked your way through this book.

Looping

Continuing with the theme of orchestral strings, it is not unusual to need a solo orchestral instrument or section to sustain for a duration longer than that of your sample. Although it can be tricky to make this sound convincing, the process of establishing the loop is not complicated, especially if you have some familiarity with the shape of waveforms.

Figure 7.8 shows the waveform of a string-section sample. As you may be able to see from the *x*-axis ruler, it lasts for slightly less than 0:05. For the sample to last longer, we need to stretch it by using a time-expansion algorithm (which may be an option), by recording a longer sample, or by finding a segment where the sound is sustaining (meaning the volume remains steady for a few seconds) and inserting loop points. Note that if there is any vibrato in the sound, be sure the end of the highlighted area will splice to the beginning in a musical fashion.

Find the command in your sample editor that defines the loop from the highlighted area—there will either be a button nearby or a menu item. By default, some editors will restrict the highlighted area's beginning and end to a location where the waveform intersects the *x*-axis—this is generally preferable as it will minimize the chance of an audible click when the loop jumps back to the beginning. Sometimes, however, the loop sounds unnatural if a segment is cut short or a vibrato pattern is altered. A solution for this is disabling the "zero-crossing" restriction and instead using a crossfade in the loop. It may take some trial and error before the results are to your liking, but keep chipping away and you should be able to get something that works to your satisfaction.

Figure 7.8
Loop Editor in EXS24. Circled on the left is the S. Loop row, and to the right can be seen the bar that represents the start and end points of the loop. The location of the loop points is easy to edit with click-and-drag. The harder part can be finding locations where the loop sounds continuous. To help, editors like the EXS24 will by default restrict placement of the start and end points to a spot where the waveform crosses the x axis—done so to avoid a click. If that still creates a jarring transition, the feature can be disabled, and you might try introducing a little crossfade as though it was an edit in an audio track.

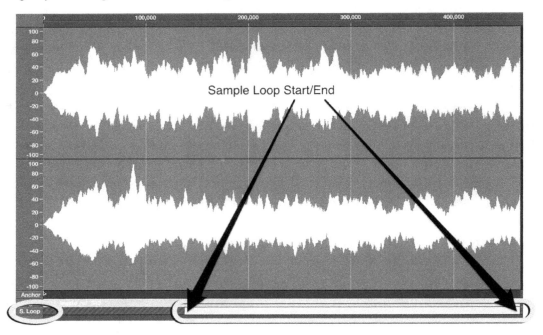

Release Samples and Control

How a sound ends is as important to its authenticity as its beginning and middle. Some samples can simply fade out or release quickly. Others have a distinctive ending that needs to be a separate part of the instrument being assembled. Pianos emit a "thunk," for example, if keys are released quickly and the damper pads strike the strings; electro-mechanical keyboards, like Wurlitzers, Rhodes, and Clavinets, do something similar. The sound of an open hi-hat, as another example, usually ends with the two halves coming together; the sound doesn't just end. The solution here is using a *release sample,* a sound that is triggered when you release the key on the keyboard. Building a sound with a well-recorded sample, a properly trimmed loop in the sustain portion, and (if appropriate) a release sample sends you in the right direction for a pretty convincing emulation of a "real" instrument (Audio Example 7.2).

Being Creative

Of course, the illusion of reality is not always the goal of a sound designer working with a sampler. Inspiring sounds can be created through manipulating beyond recognition the originally recorded sounds. But unlike those of purely synthesized sounds, the results, even after significant manipulation, will retain an organic quality if the source is an acoustic instrument or sound. A good example of this is the basis of our next Sound Design.

Sound Design 7.1—Sampled Piano Manipulation

Figure 7.9
Sound Design: altered piano.

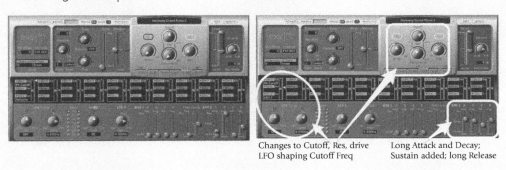

Changes to Cutoff, Res, drive   Long Attack and Decay;
LFO shaping Cutoff Freq   Sustain added; long Release

Using a sample replay instrument, open a piano that you like. Increasing the duration of the attack results in a sound that is more like a bowed string. Piano is essentially a percussion instrument, so in the amplitude envelope of the untouched version in Figure 7.9, the attack is at the fastest possible and the sustain is all the way down. Altering the duration of the decay and introducing sustain shapes the sound farther away from that of a piano, but with its timbral character still intact. Add a little release to let the sound continue for a bit after it is held. Next we play with the cutoff frequency and resonance and increase the drive for a little grit. To add a dynamic element, try LFOs tied to the cutoff frequency (as we have done here) or to the pitch or anything else you think might be interesting. The result will be familiar but hard to identify as the piano that started it all.

Digital Audio: What the Heck Is Sample Rate and What Are Bits?

Taking a sidestep for just a moment, let's get a better understanding of digital audio. For a sampler to replay a sound it must first be converted into a computer file made up of numbers that can be used later as instructions to faithfully reproduce the original waveform on demand.

We know that sound exists as a vibration of air particles varying in intensity and frequency over time. Through the use of microphones we can convert these characteristics into an alternating current (ac) electrical signal in which the amplitude of the resulting wave corresponds to the intensity fluctuations of compression (air molecules close together, high pressure) and rarefaction (air molecules pulled apart, low pressure). How quickly these variations in pressure repeat correlates to its frequency. The electrical signal generated by the microphone is an "analog" of (read: analogous to . . .) the actual sound wave. Digitizing an analog electrical signal—that is, converting a flow of ac over time to zeros and ones—can best be understood by superimposing a visual representation of a waveform onto a graph (see Figure 7.10).

Creating Sounds from Scratch

Figure 7.10
Digitizing an analog signal.

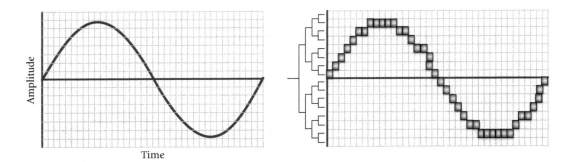

If you ever played with a connect-the-dots puzzle (drawing a line between numbered dots and revealing an image) you should have no trouble understanding at least the basics of this process. Follow the path of the sound wave in Figure 7.10 (left) and note each location where the line crosses a white block; then fill in that block (right). In essence, this is what happens in the analog-to-digital (A-to-D) converter (ADC).

The blocks going left to right represent the sample points, the resolution of which is dependent on the *sample rate* of the digital recording, and dictate the overall frequency response achievable. In theory, we need twice the number of samples for our intended frequency response, so a sample rate of 40,000 is adequate to store a range of 0–20 kHz.

The blocks going up and down correspond to the *bit rate*, which defines the dynamic range possible in our digital recording. Because each sample can be a 1 or a 0, a 4-bit recording (such as the one represented in Figure 7.10) will have sixteen possible steps of resolution ($2^4 = 16$). Because the bit rate expands exponentially, a 16-bit recording yields 65,536 (2^{16}) steps of dynamic resolution. As a rule of thumb, each additional bit adds 6 dB of dynamic range to a digital recording. And with each bit itself simply being a 0 or 1 it can exist as an electrical signal being (+) or (−) or as a reflection or no reflection from an optical disc like a CD. The higher the resolution, the more accurate the transfer to digital, the fewer errors result. Errors in digital audio translate to noise when converted back to analog because it is data that do not correlate with the rest of the signal. So how do we turn this into zeros and ones? Look at Figure 7.11.

Once again, the vertical steps represent dynamic range. In Figure 7.11 we see a 4-bit resolution with 16 steps. The horizontal steps represent sample points, the number of

Figure 7.11
Binary code from a digitized signal.

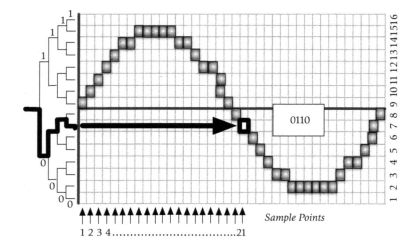

which within 1 s of time represents the *sample rate* or *sampling frequency*—shown are numbered steps up to 21. Using Step 21 as an example, we see that it falls on vertical block 7. To identify the location with a 4-bit binary code we look to the far left of the *y*-axis. There we see a single point branching into two, four times, giving us a total of sixteen steps lining up with each of our 16 vertical blocks.

If we think of this as a tree starting with a trunk on the left and branching out, we can create a path through each division to land at our targeted block. The up/down path required for reaching the highlighted position at sample point 21 is *down–up–up–down*. We could simplify that by having "down" represented by a "0" and "up" represented by a "1"—in which case the steps would be *0–1–1–0*. Therefore, our digital *word* for sample 21 is 0110. With only 4 bits of information we land in one of sixteen possible blocks. For a 16-bit tree—that would require a very large piece of paper and *lots* of patience!—you could land on one of 65,536 possible places, and with a 24-bit tree, 16,777,216!

Although there are valid arguments for recording audio at sample rates higher than 44.1 kHz, the benefit when capturing samples is minimal and, in our opinion, does not warrant the additional storage necessary for the larger sound files or the impact it will have on the sampler's demand for processing power. On the other hand, even though the vast majority of delivery formats (CD, MP3, AAC, etc.) are at the 16-bit level, there is an advantage to working at a 24-bit level. In short, the act of mixing and processing digital audio introduces complex math and rounding errors through calculations for which not all decimal places can be retained. These errors introduce noise at the lowest level of the dynamic range. If you start at the 24-bit level, you have a lot of space to work with before the noise reaches a noticeable level, and hopefully will not even reach the equivalent of a 16-bit level. Starting at the 16-bit level guarantees a finished product that will be no better—and quite possibly worse—than the 16-bit file delivered to your audience.

Hold on a second . . . after being digitized, the waveform doesn't look quite like the original. Rather than a nice, smooth waveform, it is rigid with sharp angles. This is a point of confusion for many and often, inaccurately, described as a fatal flaw of digital audio by its critics. At the risk of getting deeper into digital audio theory than is necessary for understanding its application in sampling and synthesis, suffice it to say that the nonrounded rigid angles in the reconstructed waveform (as we "connect the dots") left untouched would result in higher-frequency harmonics that were not present in the original analog signal. However, eliminating those high frequencies is relatively simple by employing a LPF during the digital-to-analog (D-to-A) stage. In all modern digital-to-analog converters (DACs, pronounced as "dacks" by those in the biz), these artifacts of the A-to-D conversion are removed and the resulting waveform is, for all practical purposes, indiscernible from the original.

Digital Audio in Practice

Having a thorough understanding of the conversion process is interesting and potentially useful, but if you are having trouble wrapping your brain around it, fear not! In this section we explore practical considerations when working with different bit rates and sample rates, and how your choices will have an impact on the resulting sound.

Creating Sounds from Scratch

Table 7.1 Dynamic Range of Common Bit Rates

Bits	Dymanic Range (X Bits • 6 dB)
8 bits	56 dB
16 bits	96 dB
24 bits	144 dB

Are more bits and samples always better? Within a reasonable limit, yes. With more bits there are fewer errors that lead to noise. Less noise means a wider dynamic range of usable information. With each bit comes an additional 6 dB of dynamic range. Although there are exceptions, 60 dB of dynamic range—from the loudest sound down to the softest—is acceptable. Looking at Table 7.1 you can see that >60 dB of dynamic range are possible with a 16-bit sound file.

So, why are 24 bits popular for recording and playback if the 96-dB range achieved with 16 bits should be more than ample? The process of mixing audio and applying effects in the digital domain is, at its essence, math. And as we know from even simple multiplication and division, sometimes the solutions are messy. For example, if you are working with a method that does not accommodate decimals, rounding to the nearest integer value will be necessary: (a) $10 \div 2 = 5$ (nice and clean), (b) $10 \div 3 = 3.333\ldots$ (not so tidy). Using (a) in a later calculation will be fine—however, we would need to use "3" and not "3.333 . . ." as the answer for (b) because our system won't accommodate the extra decimal places, so we will introduce an error into the calculations that follow.

24 bits may seem a bit extreme when you learn that even the best analog equipment will provide a dynamic range of only around 130 dB—why bother going farther? It is true that a good 24-bit recording can exceed the dynamic range of an analog signal. And if that were the end of our work, 16 bits would be adequate for any music application. However, the likelihood of further processing of the sound in the creation of a sample and/or manipulation of a mix that includes the sample (volume, pan, EQ, reverb, etc.) justifies the additional storage space of a 24-bit file to allow for the inevitable addition of noise at a level lower than will be audible.

How about sample rate? We need two samples within each wave cycle in order for it to be stored and reproduced properly. If we break a complex wave into its constituent sine waves, we can see that storing the intensity of a sine wave's compression as one sample and the intensity of the rarefaction as another we can reconstruct those values over those two samples. Of course, it won't have the nice rounded shape of a sine wave; it will look more like a square wave with the measured intensity of the compression lasting for one sample and the intensity of the rarefaction lasting over the second—basically we have turned our sine wave into a square wave. Using a LPF, though, we can shave off the unnecessary overtones and be left with just the original sine wave. Because the human hearing range is 20–20,000 Hz, we need to have a sample rate of at least 40,000 samples per second (also labeled as Hz). If we use a sample rate of 20,000 Hz we will record only up to 10,000 Hz of audio (see Table 7.2).

You are probably noticing that all of these common sample rates are able to capture a frequency greater than 20,000—so what gives? There is a benefit to the high sample rates that is different than the advantage of high bit rates. The full explanation gets a bit

Table 7.2 Resulting Frequency Response for Given Sample Rate

Sample Rate (Hz)	Audible Frequency Range (Hz)
44,100	22,050
48,000	24,000
88,200	44,100
96,000	48,000
192,000	96,000

more complex, but suffice it to say that if it is of interest, record samples for yourself at different sample rates to see if you hear a difference, and, if you do, whether the improvement warrants additional space and processing. As a recommendation for most projects, 44,100 or 48,000 Hz will be more than sufficient. ≥88,200 can be beneficial in the recording stage but we would recommend using a high-quality sample–rate converter (available in all major DAWs) to bring it down to 44,100 or 48,000 Hz when making a sample.

File Sizes

Besides the audible considerations, a real practical concern is how the bit rate and sample rate you choose affect the file size of the sample (see Table 7.3). This was of greater concern when samplers were dedicated hardware boxes with limited memory and storage. All professional software running on a Mac or PC meeting the RAM and hard-drive recommendations of the developer will accommodate high-quality samples. The more likely limit will be with your computer's processor reproducing these samples on demand: The higher the quality and the longer the soundfile's length, the greater the demand placed on the processor.

It has been said that a little knowledge can be a dangerous thing, and we are only scratching the surface of digital audio here, but we are confident this information will help you make the best choices when working with digital audio files. A deeper exploration of the digital encoding process is beyond the scope of this book but information on digital audio is widely available in other books, magazine articles, and online.

The Producer Point of View on Sampling

Samplers are, most likely, the one type of devices that have had the biggest impact on a wide variety of styles in contemporary music production. From hip-hop to film scoring, from fusion to pop, the sampler has always lived a dual personality. On one side it has

Table 7.3 File Sizes for Common Bit and Sample Rates

Sample Rate	Bit Rate	Mono	Stereo
44,100	16 bits	5 MB/minute	10 MB/minute
44,100	24 bits	7.5 MB/minute	15 MB/minute
48,000	16 bits	5.5 MB/minute	11 MB/minute
48,000	24 bits	8 MB/minute	16 MB/minute
96,000	24 bits	16.5 MB/minute	33 MB/minute

Creating Sounds from Scratch

Figure 7.12 The Emulator 1 released in 1981 by E-mu.

been (and still is) the king of acoustic sound sequencing; on the other side it has also represented the favorite tool of experimental sound designers and producers. It is this duality of roles that makes sampling such a powerful and versatile tool. In the second part of this chapter we discuss both roles. First we learn how to use the sampler as a tool for creating and sequencing realistic acoustic patches; then we unleash our creativity and learn how to use it as a more creative sound design tool.

Among the sampler synthesizers that made history in contemporary music production, there are a few that stand out for their set of features and their "relative" affordability such as the Emulator I (1981) (see Figure 7.12), the Kurzweil 250 (1984), Emulator II (1984), the Ensoniq Mirage (1984), the Akai S-900 (1986), the Casio FX-1 (1987), the Akai MPC60 (1988), and the Roland S-770 (1989). These were very powerful machines that allowed the composer and producer to achieve new levels of flexibility in sound design and production.

These early machines were particularly effective for sequencing parts of acoustic instruments such as pianos, harps, woodwinds, strings, and choirs. Their synthesis engines, though, allowed them also to create some interesting slow pads by combining sampled waves with more traditional subtractive waveforms. One area of production where samplers have been always popular and successful has been drums sounds. The ability to record our own waveforms provides the ideal tool for creating realistic and electronic drum kits by using both acoustic sounds and more creative sonorities. In this category one of the most successful drum samplers of all time was the Akai MPC series, which started with the release of the legendary MPC60 in 1988 (Figure 7.13).

The MPC combined a 12-bit sampling at 40 KHz (very respectable for the time) and a built-in sequencer with ninety-nine tracks and ninety-nine patterns.

The Sampler for Music Production and for Sound Creation

Because a sampler is probably one of the most versatile types of production tool available, in the second half of this chapter we learn some of it most important applications and uses. First we learn how to create an original drum library from scratch. Then we cover the intricacies of sequencing a convincing acoustic drum part with the main goal to make it sound as realistic as possible. Then we move to the more creative aspect of using a sampler for music production, creating a "Beat Box" kit and then creating some completely original and innovative patches. We can guarantee you that it will be an exciting and intriguing journey into sound design and production!

It is important to select the right software tools for our task. For creating a sample library we need first a DAW or digital recorder to record our samples. There are several

Figure 7.13
The Akai MPC60 sampler/drum machine by Akai.

options in this category but our favorite is Pro Tools. It has excellent audio editing and handling features, and it allows us not only to record the sample but also to edit them and preprocess them with ease and precision. If you prefer not to go the DAW route you can use as effectively a two-track audio recorder like Steinberg Wavelab, which has the advantage of providing batch processing and native sampler support.

Creating an Acoustic Drum Library

To create an original drum library, the main aspect to remember is that organization is the key word. This is true for any project that involves a large number of samples that need to be precisely catalogued and accurately placed during each step of the production process: recording, editing, mapping, and patch programming. So our first step toward creating our very own drum library is going to be the creation of a spreadsheet for collecting all the important data related to our samples. In Figure 7.14 you can see an example of the table Andrea Pejrolo used to collect the data for his *iAcoustica* drum library that he recorded, produced, and programmed for a variety of software samplers.

As you can see in Figure 7.14, in the first column we listed the name of the kit (Drum Kit A), because we had planned to sample more than one kit, and the name of the pieces of the kit we planned to record: Hi Tom, Mid Tom, Low Tom, and ride (open and roll). Note how for each piece of the kit the library has options for sticks, brushes, mallets, and wooden brushes. In addition, a critical part of creating a realistic sound library is the fact that we record each piece of the kit at different dynamics; therefore you can see how we planned to have three dynamics, *p*, *mf*, and *f* for each piece. Naming the samples correctly

Figure 7.14 A spreadsheet used to catalog the sample collection during the recording stage.

Drum Kit A		Date:		Studio:		Engineer:	Producer:
						Taroon Bali	Andrea Pejrolo
		Sticks	Brushes	Mallet		Wooden Brushes	Comments
	Hi Tom p	A-HT_St_p.aif	A-HT_Br_p.aif	A-HT_Ml_p.aif		A-HT_Wb_p.aif	
	Hi Tom mf	A-HT_St_mf.aif	A-HT_Br_mf.aif	A-HT_Ml_mf.aif		A-HT_Wb_mf.aif	
	Hi Tom f	A-HT_St_f.aif	A-HT_Br_f.aif	A-HT_Ml_f.aif		A-HT_Wb_f.aif	
	Mid Tom p	A-MT_St_p.aif	A-MT_Br_p.aif	A-MT_Ml_p.aif		A-MT_Wb_p.aif	
	Mid Tom mf	A-MT_St_mf.aif	A-MT_Br_mf.aif	A-MT_Ml-mf.aif		A-MT_Wb_mf.aif	
	Mid Tom f	A-MT_St_f.aif	A-MT_Br_f.aif	A-MT_Ml_f.aif		A-MT_Wb_f.aif	
	Low Tom p	A-LT_St_p.aif	A-LT_Br_p.aif	A-LT_Ml_p.aif		A-LT_Wb_p.aif	
	Low Tom mf	A-LT_St_mf.aif	A-LT_Br_mf.aif	A-LT_Ml_mf.aif		A-LT_Wb_mf.aif	
	Low Tom f	A-LT_St_f.aif	A-LT_Br_f.aif	A-LT_Ml_f.aif		A-LT_Wb_f.aif	
	Ride p	A-Ride_St_p.aif	A-Ride_Br_p.aif	A-Ride_Ml_p.aif		A-Ride_Wb_p.aif	
	Ride mf	A-Ride_St_mf.aif	A-Ride_Br_mf.aif	A-Ride_Ml_mf.aif		A-Ride_Wb_mf.aif	
	Ride f	A-Ride_St_f.aif	A-Ride_Br_f.aif	A-Ride_Ml_f.aif		A-Ride_Wb_f.aif	
	Ride Roll p						
	Ride Roll mf			A-Ride_Roll_Ml_f.aif			
	Ride Roll f						

and in a consistent way is absolutely crucial for a project that might potentially have thousands of samples. Let's take a look at the naming system we used for the *iAcoustica* samples. For the first sample "*A-HT_St_p.aif*," for example, we used the following convention:

Drum kit: *A*
Piece of the kit: *HT* (Hit Tom)
Type of sticks or brushes: *St* (Sticks)
Dynamic level of the hit: *p* (piano)

All the other samples of the kits were catalogued with this naming system, making it very easy for us to go through all the samples during the programming of the library. If the library is for a pitched instrument, let's say a violin for example, you should also add information about the pitch of each note/sample. This can be done by using either the note/octave or the MIDI note number systems. Adding the pitch information will make it easier during the programming stage to automatically place the right sample to the correct keyboard key. Before recording any sample, plan well in advance all the details of the recording session. In most cases you have one shot to get all the material you will need to program the library (unless you use your own facility/studio). Make a list of every single sample you plan to record, then place it in the spreadsheet we discussed previously. At the recording session check off every sample that is successfully recorded and write down any comment that you might find useful at the editing, production, and programming stages.

Understanding, Recording, and Managing Multisamples

What makes an acoustic library realistic and successful is that fact that the waveforms are not synthesized but are actually sampled from the original acoustic source. It is important to understand, though, that if we were recording only one sample per pitch or per piece of the drum kit, we would limit drastically the realism of our final result. This is because what makes a real instrument sound . . . real is the fact that every time we strike a snare drum, for example, each hit will sound slightly different from the previous one, mainly because we might hit the surface of the snare with different strengths each time. When we play an instrument at different dynamics, we will get extremely different waveforms, colors, and timbres. If we were to render all these nuances by using only one sample, we would end up with a very static and unrealistic patch. This is why we need to use the multisample technique when recording and programming our sample library. How the samples of different dynamics are triggered can be decided at the programming stage in our sampler. Usually, for percussive sounds like a drum kit, MIDI velocity is used to trigger the different samples (Figure 7.15).

As you can see in Figure 7.15, a snare drum was sampled when played at three different dynamics: *p, mf,* and *f*. Notice how the waveform for each dynamic is different. During the programming stage we assign each sample to a different MIDI velocity range. In this example MIDI velocities between 1 and 39 will trigger the *p* sample, whereas MIDI velocities between 40 and 90 will trigger the *mf* sample, and MIDI velocities between 91 and 127 will trigger the *f* sample. Listen to Audio Examples 7.3 through 7.5 to listen to samples of a snare drum played with sticks at *p, mf,* and *f* dynamics, respectively.

This technique guarantees realistic sample libraries. Of course, the higher the number of the samples at different dynamics the more realistic the final result will be. Really advanced libraries can have more than 100 different samples for each note!

The recording session for your multisample library needs to be very well organized and efficient. For large libraries we recommend having a recording engineer or at least

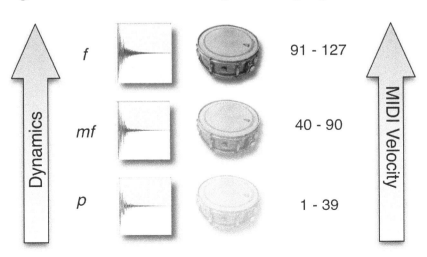

Figure 7.15
Example of a snare drum sampled at three different dynamics: *p, mf,* and *f*.

an assistant engineer who can take care of the pure technical and recording aspects of the session such as mic positioning, balance, gain setting, and so forth. Having another person on the session will allow you to focus on some keys aspects such as keeping track of the samples recorded, making sure that you get the precise sound, tone, color, and dynamic for each sample, and interact with the player or players to make sure that the performance is consistent and realistic. If you do your homework well during the recording session then the programming stage will be much easier. Make sure also that the microphone selection and placement are serving your purpose and goals. The sound of each sample needs to be captured accurately and with the right balance of presence, natural coloration, and ambience. In general, for a drum library, we recommend recording the pieces of the kit fairly dry. This option will give us more flexibility at the programming stage. As a producer of the library it is your responsibility to document the session in detail. We suggest taking notes, pictures, and/or video recordings of the session in general and of the microphone placement in particular. In case you have to go back into the studio to rerecord some of the samples, a highly detailed log of every aspect of the session will come in handy! (See Figure 7.16.)

Figure 7.16
Some pictures taken during the recording of the samples for iAcoustica.

Preproduction

Once all the samples are recorded in your DAW of choice, it is time to start sorting through all of them and pick the ones that you will want to eventually program into your sampler. This process can be tedious but, once again, it is a crucial one. Listen carefully for any imprecision, audio glitch, or tuning problems. Don't delete the samples you are planning to discard. You might need them to create alternatives for the round-robin variations (more on this in a little bit). For each piece of the kit (or instrument) now export all the best samples you have. Export the samples as mono or stereo as needed, and make sure that when you export the samples you name them correctly (according to the same naming convention we discussed earlier). If you are planning to export several variations for each piece/dynamic you should extend the naming of each sample to include the "variation number." For example, if you want to save three different variations of the sample *Kit A, Snare Drum, Sticks*, and *mezzo forte*, the three files/variations should be named

- A_SD_St_mf_v1
- A_SD_St_mf_v2
- A_SD_St_mf_v3

where A is for the drum kit, SD is for snare drum, St is for sticks, mf is for *mezzo forte*, and V1 is for the version number.

We also recommend ranking the versions in terms of preference and overall quality of the sample, so V1 would be the best sample, V2 the second best, and so on. Make sure to export your samples in an organized folders structure with all the samples belonging to the same piece of the kit or instrument in a separate folder (Figure 7.17).

Notice how we nested all the samples based on their kit, piece, and type of stick/brushes. This will make it very easy to find your samples for the next preproduction step.

Once you have selected and exported all the best samples, it is time to move to the second step of the preproduction stage. Here's where we clean the samples up, add audio effects to improve their overall quality, and fix any other audio issues that arose during recording. For this stage we suggest starting a new project in your DAW. In the new project let's create a new audio track for each piece of the kit and for each type of stick/brush (or articulation in the case of a pitched instrument). Following the previous examples, we would create a track for the bass drum, one for the snare with sticks, one for the snare with brushes, one for the snare with mallets, and so forth (Figure 7.18).

Using this track setup allows us to apply similar effects (such as equalization, compression, etc.) to samples of the same type. This methodology speeds up the production process considerably. In Figure 7.18 notice how the samples and the tracks are named

Figure 7.17
Folder structure for exported selected samples.

Creating Sounds from Scratch

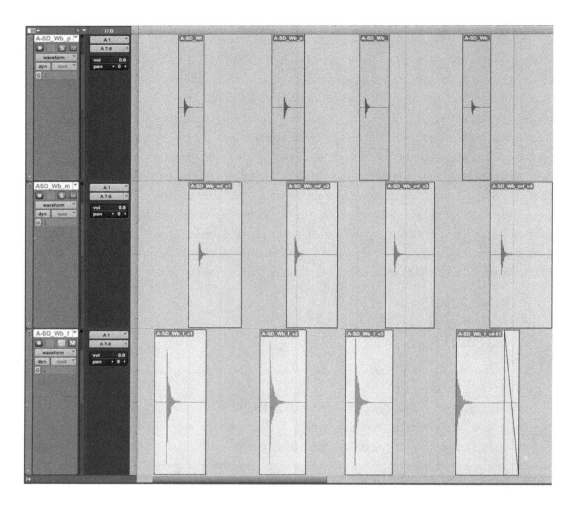

Figure 7.18
Example of project setup in Pro Tools for iAcoustica during the second stage of preproduction.

according to the naming convention we learned earlier. In this example we are looking at the three tracks for the snare played with wooden brushes with dynamics of *p, mf,* and *f*. For each dynamic we chose four variations (v1, v2, v3, v4). Listen to Audio Examples 7.6–7.8 to compare the different dynamics and variations. Once all the samples are reimported in the new projects and assigned to the respective tracks, it is time to fine-tune each sample with some basic audio editing and with some audio effects. First, we need to trim exactly the beginning and the ending of each sample to make sure that we will export exclusively the audio material that we will use in the sampler. To do so we must make sure that we leave only a few milliseconds before the beginning of the sample and that we leave a natural fade for its tail. Look at Figures 7.19 and 7.20 to compare good versus bad audio sample editing.

In Figure 7.19 we can see an example of sample start trimming. Waveforms *a* and *b* are not correct. In *a* the sample is trimmed too much into the waveform. This would lead to clicks during playback, and also we would lose part of the attack of the sample. Although waveform *b* could be ok, it is not recommendable. Remember that we can always trim the sample inside our sampler during the programming stage, but that would require extra work and also would add to the space required to store the samples on the HD. The start point of waveform *c* is perfectly trimmed. It leaves just a few samples before the beginning of the sound so that we can both have a responsive trigger and we preserve

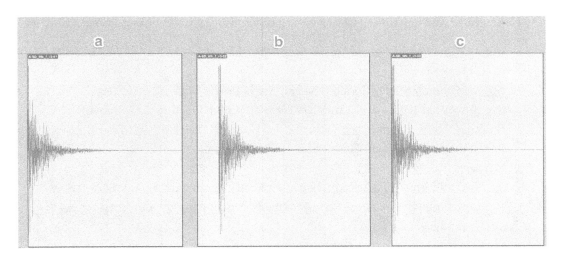

Figure 7.19
Examples of audio trimming for the start of a sample: too short (A), too long (B), correct (C).

its full attack. Listen to Audio Examples 7.9–7.11 to hear the differences among these three waveforms.

In Figure 7.20 we have an example of trimming the tail of a sample. In *a* the tail is too short. This will translate into a sample that is truncated and that creates an unpleasant effect (Audio Example 7.12). In *b* the tail is too long, resulting in a sample larger than needed (Audio Example 7.13). In *c* we have the right trimming applied to our waveform. We recommend adding a fade at the end of the waveform in order to have a smoother ending (Audio Example 7.14).

Once every sample and variation have been edited to perfection, we can move on to applying some audio effects. If we are working on an acoustic sample library, we like to keep the use of effects to a minimum. Their purpose should be to fix or tweak the sample in order to correct problems that were introduced at the recording stage. A typical example would be to use a selective equalizer to reduce a boomy effect on the low frequencies or some excessive "boxy" artifact in the mid low range. We recommend using mainly two types of effects: equalization and compression. Remember to use them with moderation at this stage. We can always add more drastic effects during the programming of the patch in the sampler. Here we are looking for only some small adjustments in order to have all the samples that belong to the same kit/piece/dynamic as leveled and consistent as possible.

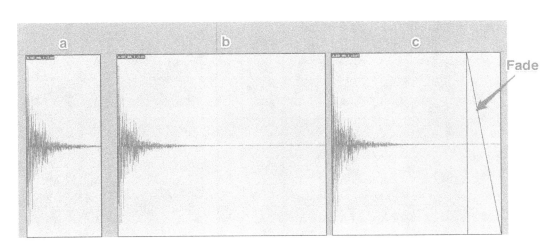

Figure 7.20
Examples of audio trimming for the tail of a sample: too short (A), too long (B), correct (C).

Use a multiband equalizer to fix common sonic issues with the samples such as these:

- <u>Reduce low rumbles or noises:</u> Reduce the gain with a LPF with a frequency set to 60 Hz or lower.
- <u>Reduce boomy effect:</u> Reduce the gain with a peak EQ with a center frequency set between 80 and 100 Hz with a high Q value (1/2 octave).
- <u>Reduce "boxy" resonances:</u> Reduce the gain with a peak EQ with a center frequency set between 250 and 800 Hz with a Q value between 1 and 2 octaves.
- <u>Reduce mid–high nasal resonances:</u> Reduce the gain with a peak EQ with a center frequency set between 700 and 1500 Hz with a Q value between 1 and 2 octaves.
- <u>Reduce dull spots:</u> Boost the gain with a peak EQ with a center frequency set between 6 and 10 kHz with a Q value between 1 and 2 octaves.
- <u>Reduce harshness:</u> Reduce the gain with a peak EQ with a center frequency set between 3 and 9 kHz with a Q value between 1 and 2 octaves.
- <u>Reduce high-frequency noise:</u> Reduce the gain with a high shelving filter with a frequency set around 12 kHz or higher.

Depending on the type of sample you are dealing with, we also recommend using a gentle and mild compressor to render more uniform the samples of the library. This holds true particularly for percussive sounds and for samples that have a sharp attack, such as the one in a drum kit. Usually we prefer using a compressor that has a vintage sonority. This will infuse back some of the acoustic feel of an analog recording. For iAcoustica, for example, Andrea used a classic BF76 from Bomb Factory (Figure 7.21).

Although it is not possible to give precise compressor settings for every sample or instrument that you plan to program, there are some general things you should keep in mind:

- Use a mild compressor *only*, unless really necessary. Remember that we can always tweak the sample during the programming phase in the sampler.
- For percussive sounds, use a slightly longer attack in order to preserve the fast transients. For sustained notes use a longer release if you want to increase the sustain of the sample.
- Always compare your sound with and without compression applied and decide if you are going in the right direction.

Figure 7.21
The BF76 Bomb Factory compressor limiter.

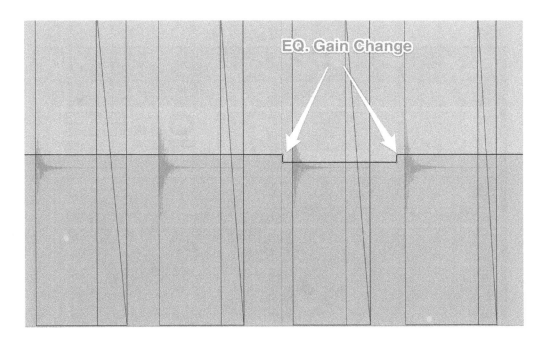

Figure 7.22
Example of automation used to change the parameters of the EQ from sample variation 2 to variation 3.

Because you are applying these effects on a whole track that includes different variations of the same sample, we recommend using automation to adjust the parameters for single samples. For example, look at Figure 7.22 where we used automation to adjust slightly the center frequency of the peak EQ for variation number 3 of the snare sample.

After finishing tuning your samples, it is now time to export them again before importing them in our sampler. Remember that the naming convention is crucial here. Use the same convention you used earlier. We recommend, if you know where you want to map the samples in your sampler, to also add the key range for each sample. As we are going to see in the next section, mapping the sample to the keyboard sampler is the next very important step that we will tackle. Although this action can be done manually, having our sampler automatically map the samples based on their names can speed up the process considerably. To do so you can add information where you would like each sample to be placed by simply adding the range in terms of starting key and ending key. For example if we know that the bass drum sample needs to be place on MIDI key C1 (that's note "C," octave "1"), we can name the sample KitA_BD_mf_C1. If we want to place a high tom from B1 to C2, we would name the sample KitA_HT_mf_B1-C2. This system is extremely useful when used for pitched instrument libraries because it allows us to automatically place the samples on the correct key according to their pitch.

Programming a Drum Library

After all this work, now it is finally the time to start working with our sampler. Although for the rest of this chapter we will use NI's Kontakt (Figure 7.23) as our programming tool, all the modern software samplers handle programming in a similar way with almost identical terminology.

Creating Sounds from Scratch

Figure 7.23
Native Instruments' Kontakt sampler.

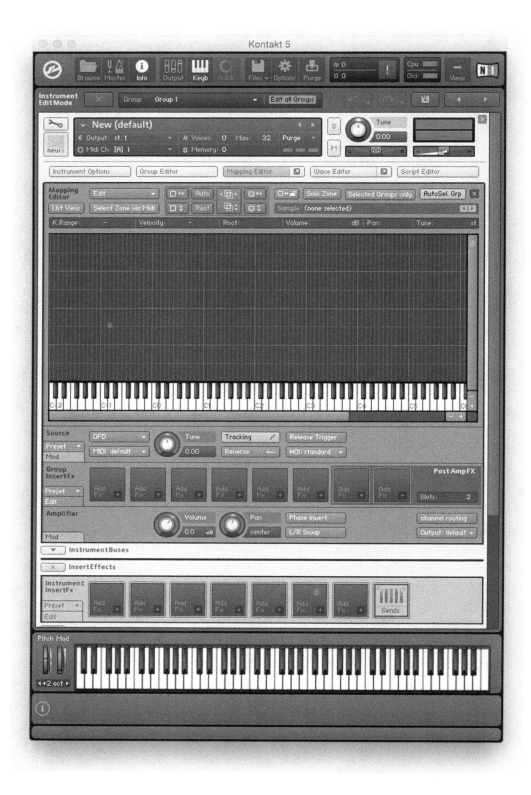

The steps involved in going from samples to a full drum library are several. To have a clear overview of the process, we listed each one with a brief description in Table 7.4.

Mapping

Depending on the number of samples we plan to have in our patch, this process can be quick or can take a good amount of time, so make sure that you leave plenty of time for

Table 7.4 The Steps Involved in Programming a Drum Library

Step	Description	Comments
1	Mapping	During the mapping process we assign the samples to the keys taking care also of the assignments of different samples for different dynamics.
2	Sample editing & looping	After mapping, we make sure that all the samples are perfectly edited (in terms of start and ending). We also work on the looping of the samples if necessary.
3	Multisample programming	If we are dealing with a multisample patch, we need to make sure to program MIDI velocity trigger zones and crossfades between samples.
4	Group editing	Then we move to editing the groups in order to easily edit parameters of samples that are similar or that belong to the same category (for examples all the Hi-Hat samples in a drum kit).
5	Envelopes and modulation programming	In this step we set up all the modulators and envelops including a velocity-controlled amplifier to create a responsive and dynamic patch.
6	Effects	In this stage we add effects such as reverb, ambience, group-based equalization etc.

this process. The mapping process involves the assignment of each sample to a key or key range. In the case of a drum library we are going to use the traditional General MIDI mapping standard (Figure 7.24), which will guarantee you maximum compatibility with other DAWs and notation software.

This is how you should import and map your samples in Kontact (Figure 7.25):

1. Launch Kontakt and create a new instrument by selecting Files>New Instruments.
2. Click on the wrench tool to enter the Instrument Edit Mode
3. Click the Mapping Editor tab.

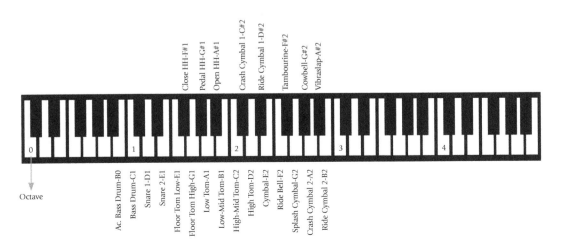

Figure 7.24

The drum mapping in the General MIDI standard.

Creating Sounds from Scratch

Figure 7.25
Creating a new instrument and opening the Mapping Editor in Kontakt.

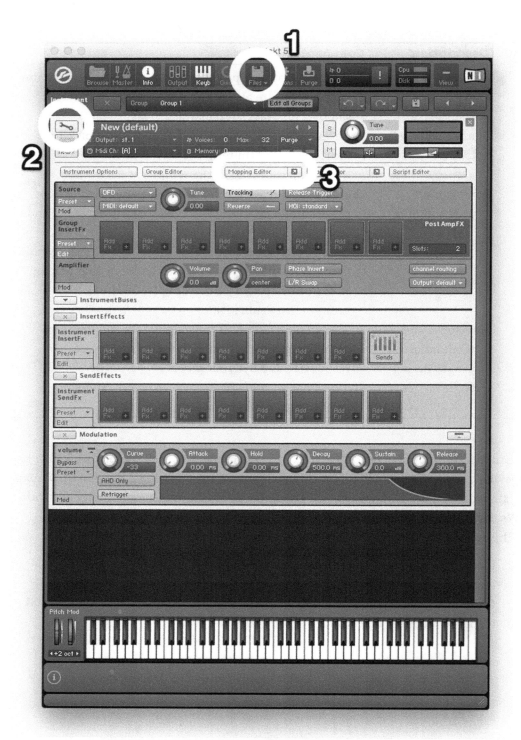

Once the mapping editor is open, you can simply drag your samples into it (Figure 7.26). As you can see, the sample at the moment is assigned to the keys sequentially. You can move each zone (a zone is a combination of samples assigned to a key or key range) manually to follow the General MIDI map we saw earlier.

If your sampler has the "Auto Map" feature we described earlier, you can automatically map your samples to a specific key, or key range, based on the information preset

Sample-Based Synthesis

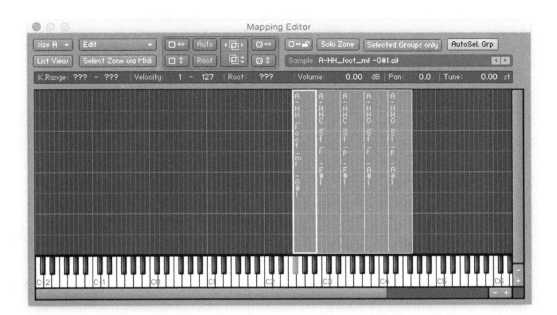

Figure 7.26
Mapping samples to keys.

in the name. In Kontakt we can easily do so by first setting up the Zone Auto Mapping parameters (Figure 7.27), where we set the rules for linking naming information to parameters of the zone.

To do so select a zone (any zone will do it), then right-click on the selected zone and select "Auto Map - Setup." Now set the rules for the automatic mapping. In this case we have set the last part of the sample names to be "set to single Key." This means that when the sampler reads the file name and detects a note name (for example F#1) it will automatically map that sample to that key. If you need to have a sample mapped to a range of key, make sure to include in its name the range, for example "F#1–G#1." Once we set up the rules we can click the "Apply" button. If all the other samples imported follow the same naming system you can select all the zones and auto map them at the same time by selecting them all, right-clicking on them, and selecting "Auto map functions>Auto map selected," as shown in Figure 7.28.

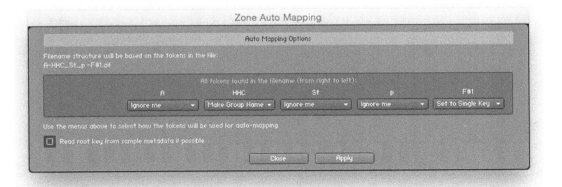

Figure 7.27
Zone Auto Mapping parameters.

Creating Sounds from Scratch

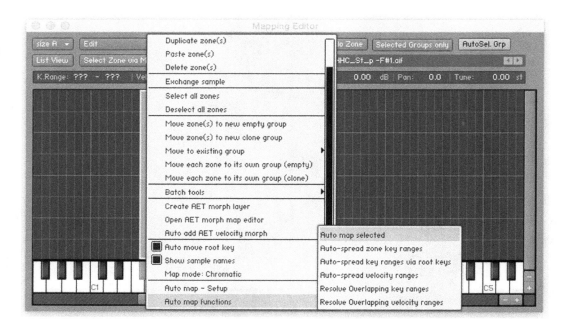

Figure 7.28
Auto mapping several samples at the same time.

Once you apply the mapping to all of the imported samples they will be moved to the correct key, as shown in Figure 7.29.

If you noticed, so far we have been dealing with only the mapping of samples to keys but we have not yet dealt with the different dynamics for each sample. We have imported and assigned the Hi-Hat HH samples for both *p* and *f* to the same key and with the same MIDI velocity range. Of course we want to have the *p* sample be triggered by lower velocities and the *f* sample by higher velocities to re-create a real acoustic effect. To do so we need to select each dynamic and assign them to a velocity range. In Kontakt that can be easily achieved, as follows (Figure 7.30):

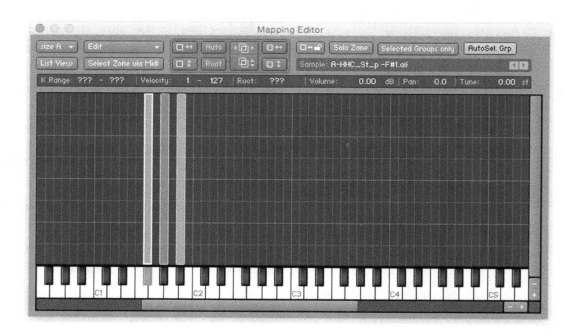

Figure 7.29
The HH (closed, foot, open) samples assigned respectively to F#1, G#1, and A#1.

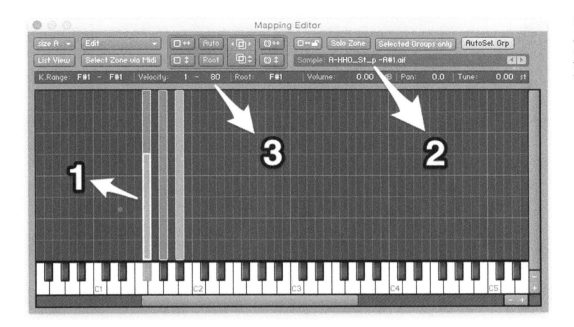

Figure 7.30
Assigning different dynamics to different MIDI velocity ranges.

1. Select the samples belonging to the same key or key zone.
2. Then make sure to select the first dynamic (*p* in this case).
3. Finally, set its MIDI velocity range.

As you can see in Figure 7.30, we set the velocity range for the closed HH at a *p* dynamic from 1 to 80. Now you can repeat the same procedure for the other dynamics of the same piece of the kit. In this case, for example, we are going to set the velocity range of the *f* sample of the closed HH from 50 to 127 (Figure 7.31).

As you can see in Figure 7.31, the two zones of the HH closed with dynamics *p* and *f* are overlapping over MIDI velocities 50 to 80. Although it is possible to have the two zones not overlap, when we have different dynamics for the same instrument or piece of the kit, we highly recommend having an overlap. This will guarantee a smoother transition between the *piano* and the *forte* samples. Although at the moment, if you play a MIDI note of F# with velocities between 50 and 80 you will hear both samples,

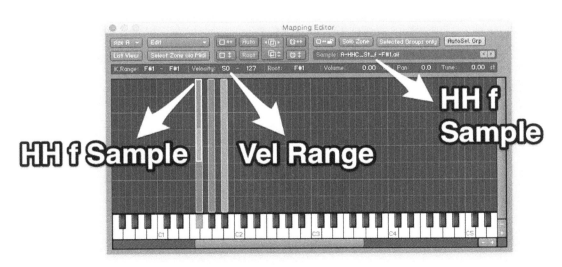

Figure 7.31
Assigning the MIDI velocity range of the second dynamic/sample.

Creating Sounds from Scratch

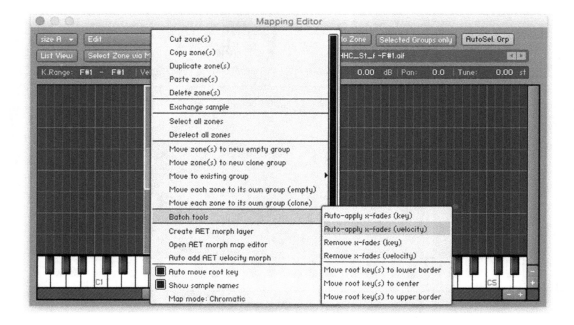

Figure 7.32 Crossfading between two zones that have overlapping velocities.

we are going to learn how to have the two zones crossfade into each other in order to have a smoother and more realistic transition between the two dynamics. To create a crossfade between two zones in Kontakt, select both zones, right-click on the highlighted zones and select "Batch tool>Auto-apply x-fades (velocity)" from the drop-down menu (Figure 7.32).

The result will be a smooth transition between the two zones in the velocity range in which they overlap (Figure 7.33).

Notice how in Figure 7.33 the samples fade into each other for the closed HH (F#1) and the open HH (A#1). For the Foot HH we have only one sample, so no crossfade is necessary.

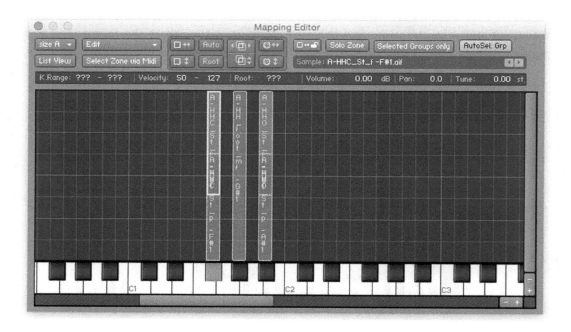

Figure 7.33 Crossfade applied to two overlapping zones.

Sample-Based Synthesis

Figure 7.34
Sample/region parameters editable from the Mapping Editor.

Fine-Tuning the Samples

When mapping the samples often we realize that some extra tweaking is necessary. There are two main ways to fine-tune our samples: The first (and more immediate way) can be done right from the Mapping editor (Figure 7.34). From here we have quick access to the most basic parameters of any selected sample/region such as key range, velocity range, root key, volume, pan, and tune. Use these parameters to apply quick fixes. We find these parameters extremely useful.

If we need to edit more advanced parameters such as start and end of the sample, loop points, fades, normalization, and so forth, we can us the built-in wave editor (Figure 7.35).

In theory we should have taken care of any sample editing in the previous phases, when we were editing and then exporting the samples from our DAW. Nevertheless this is a handy editor for last-minute fixes.

Working With Groups

As we learned in the first part of this chapter, in a sampler we can assign two or more samples/zones to a group, therefore allowing us to apply parameters or effects to all the samples/zones belonging to the same group. This is really useful, particularly when we are dealing with libraries that have hundreds of zones. For the purpose of building a drum patch, the group function comes in handy to simulate how a real HH works. When we play a real HH in open position and then we switch to the closed position or foot, the open sounds stops ringing. In a sampler, if we play the open sample and then the closed sample, or the foot HH sample, the open sample will keep ringing, resulting in an artificial sound.

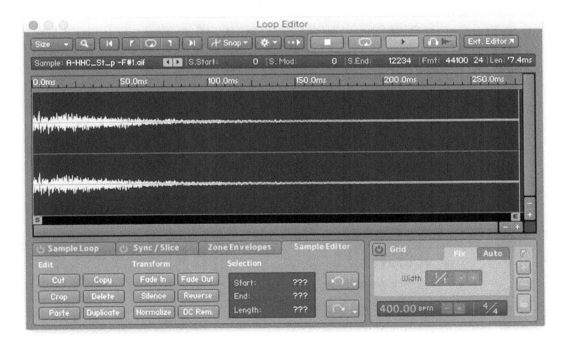

Figure 7.35
The waveform editor in Kontakt.

Creating Sounds from Scratch

To reproduce the realistic HH response in our sampler, we need to use our sampler's group functions and their parameters. For example, in Kontakt we can assign different samples to the same group (let's say the HH group) and then decide that the HH group has a polyphony of only one voice. This means that any sample that belongs to the HH group will be able to have only one active sounding voice at the same time. If the open HH sample is sounding, it will automatically stop when another sample, belonging to the HH group, is triggered (for example, the closed HH sample). To achieve this in Kontakt we need first to create a new group with all our HH zones. To do so we follow these steps (Figure 7.36):

1. Select all the HH zones.
2. Open the Group Editor.
3. Right-click on a selected zone, and from the drop-down menu select "Move Zone(s) to new empty group."
4. A new group will be created in the Group Editor.

Figure 7.36

Assigning different zones to a group in Kontakt.

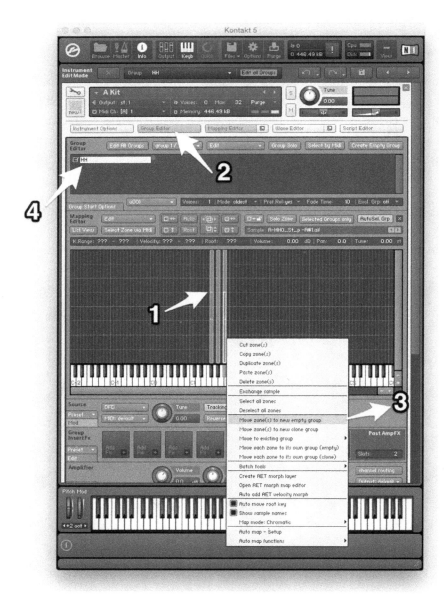

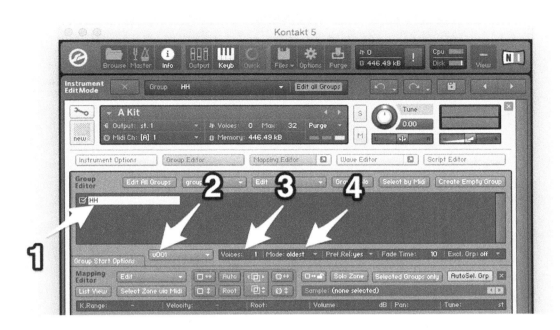

Figure 7.37
Setting the voice polyphony for the HH group in Kontakt.

Once we have all the HH samples grouped, we need to edit the parameters of the HH group. In particular, we want to make sure that the HH group has only one voice of polyphony. To do so we follow these steps (Figure 7.37):

1. We select the group we want to assign the voice polyphony to (in this case the HH group).
2. We assign a Voice Group (1–128) to the select group. Voice Groups are used to assign the voice polyphony parameter.
3. We set the "Voices" to 1, which means that only one note at the time can be played for zones belonging to the selected group.
4. We set the Mode to "Oldest," which means that the latest zone triggered will be the one taking over the single available voice in the group.

Now we are able to play only one HH zone at this time, making our "virtual" sampled HH behave like a real one!

At this point we can start importing the other samples of the drum kit and repeat the previous steps for each piece of the kit.

A finished kit would look like the one shown in Figure 7.38.

Programming Envelopes and Modulations

Now that we have mapped our samples to the keyboard, programmed the crossfades between different dynamics, and set up the groups and voice polyphony, we have a functioning drum kit. But we are not done yet! The next crucial programming step is the addition of modulations and envelopes. As you know by now, these are crucial aspects of the sound-shaping experience. For our drum kit, we need to first make sure that we can control the volume of the samples by using MIDI velocity. If this setting was not in place

Creating Sounds from Scratch

Figure 7.38
The iAcoustic Kit A with sticks.

we would not be able to control the dynamics of our parts. At the moment we can control only the sample switching, as we discussed earlier in this chapter, but we also need to have full control over the volume of the samples. Although this function can be assigned to a long list of control parameters, such as MIDI CC, aftertouch, and so forth, for percussive sounds we need to use MIDI velocity to guarantee the same response that we experience when playing an acoustic kit. In Kontakt, modulators have basically two parameters, a controller (i.e., a MIDI CC, a LFO, a MIDI message, etc.) and a destination, such as the amplifier (volume), pan, and so forth. To add a modulator for the amplifier that controls volume, we follow these two steps (Figure 7.39):

1. We click on the "Add modulator . . ." tab.
2. We select the source of the modulation (in this case we are going to select "External Sources>velocity").

A new modulator will be inserted. As you can see in Figure 7.40, now Velocity (1) controls Volume (2). To adjust the amount of control applied by the source to the destination, you can use the Modulation Intensity slider (3). Use this last parameter to make the modulation control less sensitive (move the slider to the left) or more sensitive (move it to the right).

We can follow the same procedure to set up an envelope for the amplifier. Under the tab "Add modulator," select "Envelopes>AHDSR" as the source and select "Volume" as

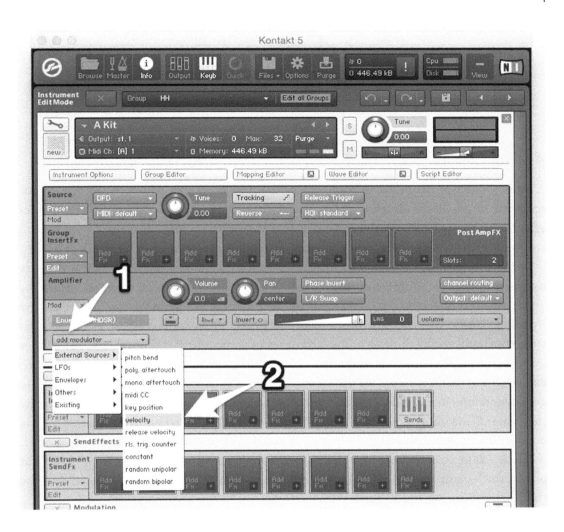

Figure 7.39
Adding a modulator in Kontakt.

the destination. The new modulator will be shown at the bottom of Kontakt, as shown in Figure 7.41.

Although for a drum kit we are not going to alter the Amp Envelope, we will be definitely using it for any other sustained sound that we will create.

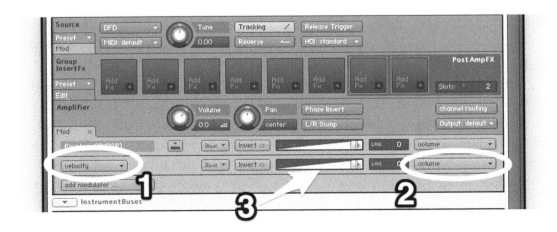

Figure 7.40
Controlling volume via MIDI velocity.

Figure 7.41

The envelope of the Amplifier in Kontakt.

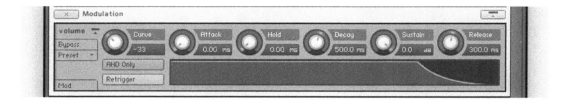

Adding Effects

At this point our drum kit is ready. As a final touch we can add some effects to fix, enhance, or creatively alter the patch. In Kontakt we have access to a good selection of effects that range from reverbs and delay, to distortion and amp simulators, to compressors and modulation effects. When it comes to adding effects to a sample patch or library, we have the options of using the effects as either insert or send. The main difference between these two approaches resides in the fact that if we use an effect as insert the entire signal is affected by it, whereas if we use an effect as a send we can decide, via the "Auxiliary Send" control, how much of the original dry signal is sent to the effect, and therefore is altered by the effect. The advantage of using a send is that we can "share" that effect among different zones or groups. Usually we use sends for time-based effects such as reverb and delay, whereas we use inserts for effects such as compression, equalization, and distortion. In Kontakt we also have the possibility of adding effects to the entire instrument (meaning to all the zones/groups together) or on a per-group basis. This second option is particularly useful when we want to apply specific effects to specific groups. A classic example is the one in which we want to add a reverb on the snare and HH but not on the bass drum. To add an insert effect to the entire instrument in Kontakt we use the "Instrument Insert FX" section (Figure 7.42).

We can have up to eight effects inserted. To add an effect to the chain simply click on one of the open slots and choose the desired effect. In Figure 7.42, for example, we inserted a three-band EQ in the first slot and we are going to add a compressor in the second slot.

To add a send effect to the entire instrument, we follow these steps (Figure 7.43):

1. We need to first add the effect in the "Instrument SendFX" section.
2. Then we add a "Send Level" instance on one of the insert slots.
3. Finally we use the send level knob to adjust the amount of dry signal sent to the effect

In the previous example we added a convolution reverb as an Aux send on the entire instrument.

Although the use of effects at the instrument level can work in some cases, we prefer being in control of the effects for each group. We might, for example, use a different EQ or compressor settings for different zones, or even apply reverb or delay only to certain groups. A typical application of this technique would be when we are working with a percussion instrument. In general, we would apply reverb on the snare and HH but not on the bass drum. Kontakt allows us to have different inserts and send levels for each

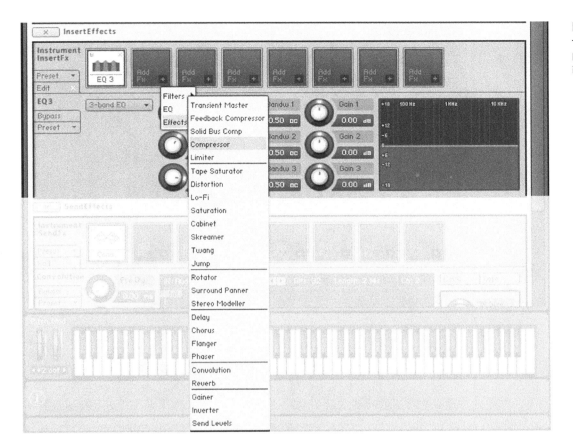

Figure 7.42
The "Instrument Insert FX" section in Kontakt.

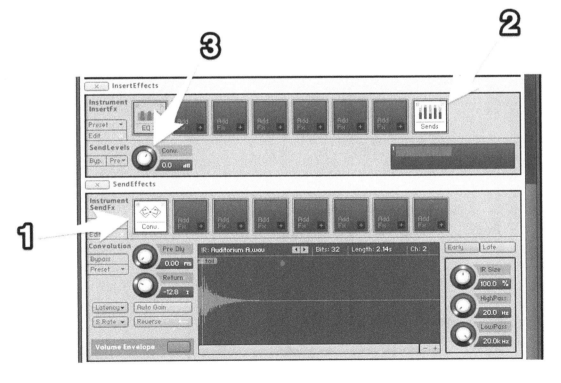

Figure 7.43
The "Instrument Send FX" section in Kontakt.

Figure 7.44
Applying effects to a specific group in Kontakt.

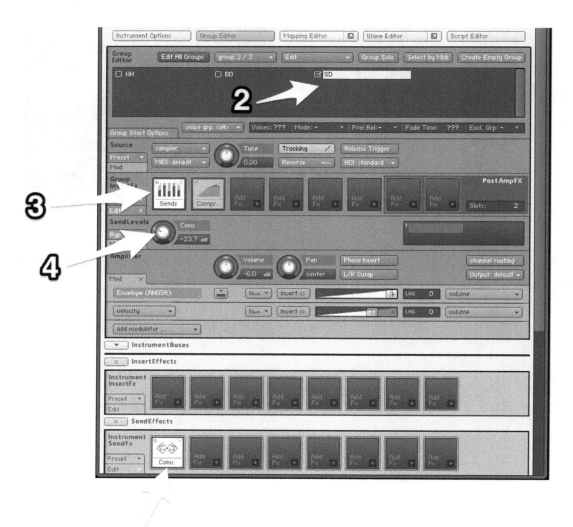

group. To apply, for example, a reverb to our snare group only, we would follow these steps (Figure 7.44):

1. We add a reverb effect in the "Instrument SendFX" section as we did previously.
2. We make sure that we have the right group selected (SD in this case).
3. In the "Group InsertFX" pane, we insert an instance of "send levels."
4. We use the send level knob to adjust the amount of dry signal sent to the effect.

Creating Variations Through Round-Robins and Random Sample Selection

Although so far we have been triggering always the same sample for a specific zone/dynamic, we strongly advise bringing your libraries to a higher level of realism by using random sample selection and round-robins. Even though these two techniques are similar, they are usually applied to different types of instruments or libraries. The main idea is to use variations of the main samples (remember that earlier we created four variations of each sample/dynamic?) and have the sampler switch randomly or according to a specific preestablished order, among the sample variations. This will create a sense of realism because we won't hear always the same sample when pressing a key. What is the difference between random sample selection and round-robin? In a random selection the sampler will pick randomly from among a series of variations each time the key is pressed. For example, suppose the four variations samples are called

    KitA-SD-WB-p-D1-V1
    KitA-SD-WB-p-D1-V2
    KitA-SD-WB-p-D1-V3
    KitA-SD-WB-p-D1-V4

where KitA is for drum kit A, SD is for snare drum, WB is for wooden brushes, p is for a dynamic of *piano*, D1 is the root key, and V is for the version number.

If we use a random variation alternation, every time we press the D1 key on the keyboard controller, we might get any random sequence of sample variations such as

KitA-SD-WB-p-D1-V2, KitA-SD-WB-p-D1-V4, KitA-SD-WB-p-D1-V1, KitA-SD-WB-p-D1-V2, KitA-SD-WB-p-D1-V4, KitA-SD-WB-p-D1-V2, and so forth.

As you can see, there is no way to predict which sample variation will be played next. This option is particularly effective for percussive instruments for which it is important to recreate the random effect of a stick hitting a snare or a cymbal.

In a round-robin system, on the other hand, we get to decide the order in which each variation is played. Using the four variations of the previous example, in a round-robin system we get to assign the priority order of each sample. Suppose we assign to these samples the following order:

    KitA-SD-WB-p-D1-V1 ---->3
    KitA-SD-WB-p-D1-V2 ---->1
    KitA-SD-WB-p-D1-V3 ---->4
    KitA-SD-WB-p-D1-V4 ---->2

Every time we trigger the D1 key, we will get the following sequence of samples: KitA-SD-WB-p-D1-V2, KitA-SD-WB-p-D1-V4, KitA-SD-WB-p-D1-V1, KitA-SD-WB-p-D1-V3, KitA-SD-WB-p-D1-V2, KitA-SD-WB-p-D1-V4, KitA-SD-WB-p-D1-V1, KitA-SD-WB-p-D1-V3, and so forth.

This technique is very effective when used, for example, with string instruments such as the violin to simulate up-and-down bow articulations.

In Kontakt we can easily program sample variations by using both techniques. Let's learn how to do it.

Random Sample Variation in Kontakt

To create a random variation of samples in our Kontakt instrument, we first map our sample variations to a zone (this could be the same zone or different zones, depending on our needs). For this example we mapped four different variations of a snare drum played

Creating Sounds from Scratch

Figure 7.45
The Mapping editor in List View mode with the four sample variations.

with wooden brushes at a dynamic of *mf*. Because all these samples are assigned to the same key, it is easier to look at the mapping editor using the "List View" (Figure 7.45).

Now, we select all four samples, right-click on the selection, and choose "Move each zone to its own group (empty)," as shown in Figure 7.46.

Four new groups will be created automatically, each containing a single zone. Now, to randomly switch zones, we follow these steps (Figure 7.47):

1. We select all the groups that we just created.
2. We click on "Group Start Option."
3. We select "Cycle Random."

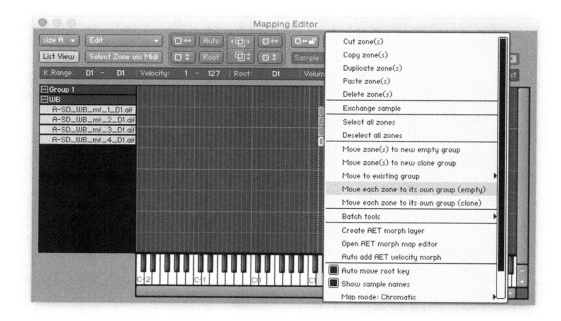

Figure 7.46
Assigning each zone to a different group.

Sample-Based Synthesis

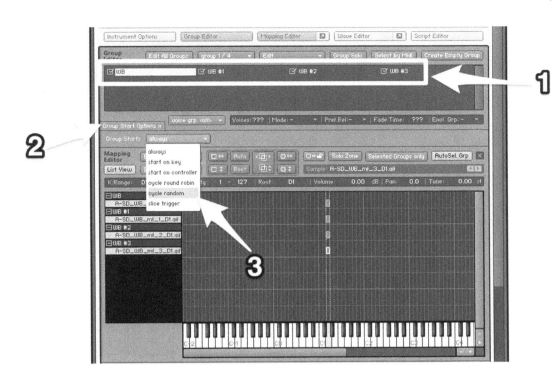

Figure 7.47
Setting the Group Start Options to random.

Every time we will be hitting the key assigned to these zones, a randomly selected zone will be played.

Round-Robin Variations in Kontakt

The procedure to assign round-robin sample selection is very similar to the one used to program random sample switching. The preparation is in fact exactly the same. First, we import and map the sample variations to one of several keys according to our needs. Then we assign each variation to a unique group. Now we need to give each group a round-robin order. To do so, we follow these steps (Figure 7.48):

1. We select the first group.
2. We click on the "Group Start Option" menu.
3. We select "Cycle round robin."
4. We set the "Position in round robin chain" to the order number we want.

Now we repeat the same steps for the other groups. In Figure 7.49 you can see how we have assigned the second group to round-robin position 2.

Tips on How to Sequence for Acoustic Drums

Because we have learned how to create an acoustic drum kit it is important to understand also the major rules for sequencing realistic and convincing acoustic drum parts. In the next few paragraphs we discuss the most important tips and tricks to bring to life MIDI drums!

One of the main rules for re-creating convincing drum parts with your DAW is to try to avoid using your keyboard as a controller and use MIDI pads instead. A major problem with using a keyboard controller to sequence drum instruments is the lack of a wide

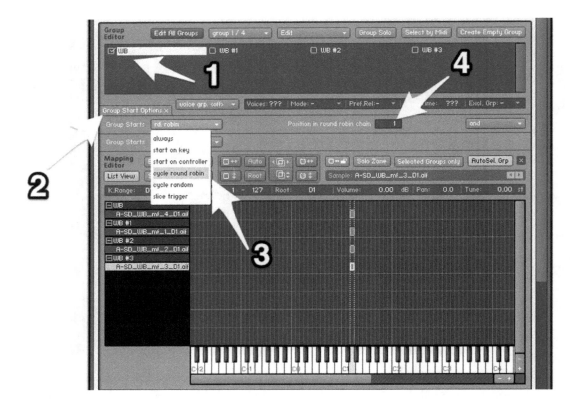

Figure 7.48
Assigning groups to a round-robin position in Kontakt.

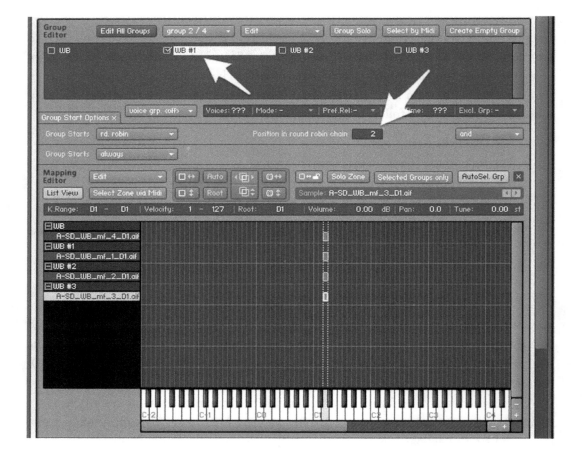

Figure 7.49
Assigning the second group to round-robin position 2.

Figure 7.50
A MIDI drum pad (Courtesy of Roland Corporation US).

dynamic range (in terms of MIDI velocity data) required for rendering drums and percussion instruments in a realistic way. The action of a keyboard gives you only more or less 1" to express 127 different velocities (from 1 to 127), making it very difficult for the performer to select the desired dynamic accurately. A MIDI pad (Figure 7.50) allows for much higher control over the dynamic response of the controller and therefore a much more detailed handling of the multisample sounds.

The percussive MIDI controllers can vary depending on their size, style, and features. Nowadays you can find three main types of percussive controllers: drums, pads, and triggers. Drum controllers re-create the entire drum kit, including bass drum and cymbals. Each pad of the kit has several velocity-sensitive zones that can be assigned to different notes and MIDI channels. When sequencing drum parts, use separate tracks for each piece of the kit. This means that, usually, toms, HH, and cymbals (sometimes it is good to separate the cymbals into separate tracks for each type such as ride, crash, etc.) will each be on their own track. We recommend putting bass drum and snare on the same MIDI track and recording them in the same pass to obtain a more natural feel. The main reason for having different tracks for each piece of the drum set is that it will give you more flexibility at the editing and quantization stage. When sequencing a drum part, start with the kick and snare; this will provide a solid foundation. Then move to the HH and cymbals, and finish with toms and fills. After each pass make sure to quantize (if necessary) the part you just laid down; otherwise you will be building each pass over tracks that are not rhythmically solid. This approach will translate into a much more cohesive and solid drum part. Avoid sequencing only four or eight bars and then copying and pasting the part over and over. Remember that for a successful and realistic MIDI rendition you have to avoid repetitions as much as possible. After you have recorded the drum part, it is time to fine-tune the quantization setting in order to make the part more human and realistic. Quantizing a MIDI part allows you to correct rhythmically the position of MIDI data in general, and notes in particular, according to a grid determined by the "quantization value." This value must be equal (or smaller) to the smallest subdivision that your drum part contains. For example, if you recorded a percussion part in which you have mixed rhythms such as eighth and sixteeth notes, then you have to set the quantization value to sixteenth notes. One of the biggest mistakes you can make when quantizing a drum part is to use full quantization (100%), also called "Straight," which will take the life out of your parts, making your sequence sound more like a drum machine than a real drummer. To avoid this effect, we recommend using advanced quantization tools such as *strength, sensitivity, and swing*. Let's take a look at how these three options work. Strength allows you to control how much the MIDI events will be quantized. If you choose a value of 100% the events will be moved all the way to the closest grid point. At the other extreme, if you choose a value of 0% their original position will not change. If you choose a 50% value, the events will be moved halfway between their original position and the closest grid point. This option gives you great control over the "stiffness" of the quantization. Although with a 100% strength (usually the default) your parts will sound very rigid and mechanic, choosing a value between 50% and 80% will help maintain the original smoothness of the parts and at the same time correct the major timing mistakes.

By controlling the sensitivity of the quantization algorithm you can choose which events will be quantized and which will be left unquantized based on their relative position to the grid. In most DAWs the sensitivity parameter (also sometimes called "Exclude within" or "Q-Range") ranges from 0% to 100% (others allow you to choose the range in ticks). With a setting of 0% no events will be quantized, whereas with a value of 100% all the events in the selected region will be affected. Any other value in between will allow you to extend or reduce the area around each grid point influenced by the quantization action. With a setting of 100% each grid point has an area of influence (a sort of magnetized area) that extends 50% before and 50% after the point (a total of 100%). Events that fall into these two areas will be quantized and move to the closest grid point. By reducing the sensitivity, the area of influence controlled by the grid points is reduced. If you choose a negative value for the sensitivity parameter you will achieve the opposite effect: Only the events that were played farther from the grid points will be quantized, leaving the ones that were fairly close to the click in their original position. This setting is perfect to fix the most obvious mistakes but leave the overall natural feel of your performance intact. On a practical level, a sensitivity value between 50% and 80% can be used to fix major rhythmic mistakes but keep the overall feel and grove of the performance intact.

Try also to use the swing option to control the rhythmic flow of your drum parts. Adding a little bit of swing percentage (usually the range is between 0% and 100%) will help loosen the grove a bit without making it swing too much.

Listen to Audio Examples 7.15 and 7.16 to compare a drum part sequenced without advanced quantization techniques and one sequenced with advanced quantization techniques, respectively. Notice how the second one sounds much more realistic.

Creating a Beatbox Drum Kit

A variation of a traditional acoustic drum kit is a Beatbox kit. Beatboxing is a vocal technique in which percussive sounds are created with only the human voice or sometimes the human body. Its origins can be tracked all the way to traditional rural music more than 100 years ago. It is hip-hop, though that brought Beatboxing to the attention of the masses in the mid-1980s with artists such as Doug E. Fresh, Rahzel, Ready Rock C, and

Buffy. From a sound design point of view, the rules for creating a Beatbox kit are much more relaxed compared with those for an acoustic kit. Besides the fact that we need the usual drum sounds such as bass drum, snare(s), HH, cymbals, and toms, the sounds range from dry and clean all the way to distorted and with a lot of effects. Creating a Beatbox kit can be a lot of fun, and it allows you to let your imagination run free! Because of its hip-hop tradition, we are going to learn how to create a Beatbox kit using the iMPC software app for iPad. iMPC is a software replica of the original MPC60 by Akai (Figure 7.51).

The steps required for building a Beatbox kit are very similar to the ones we learned when we created the acoustic kit. First we need to record the samples we will use for our kit. Whereas for the acoustic kit we were tied to recording the pieces of the kit in a very realistic way, for a Beatbox we are free to interpret each sound as we like. The most important thing to remember is that we need the basic drum sounds such as bass drum, snare, and HH (open and closed) to start with. Let's record these samples first. After launching the iMPC app start a new empty project, then navigate to the Program page. Here's where we can record new samples, edit them, and assign them to a matrix of 16 x 4 pads (Figure 7.51). To record a new sample, tap on the pad you would like to assign the sample to, for example pad A01, and then tap on "New Sound." This will bring up the Record page (Figure 7.52).

To record our sample, we can use the built-in mic of the iPad, any input of a connected audio interface (Mic/Line), a section of any song loaded in our iTunes library (Turntable), or another sample from a different app (Inter-APP), or we can resample an existing sample. Because the fun of creating a Beatbox is to perform our own samples we will be using the Mic/Line option. An important thing to keep in mind is that traditionally Beatbox

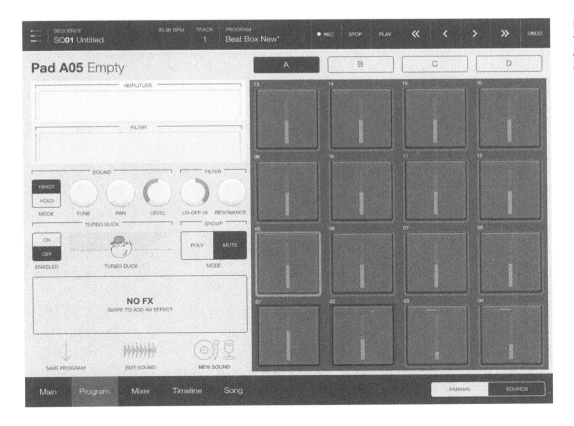

Figure 7.51
The iMPC pro app running on an iPad.

Creating Sounds from Scratch

Figure 7.52
The Record page in iMPC Pro.

samples are not exclusively about quality of the recording but are more about instinct and spontaneity. This is why we often like to use either a low-quality microphone or even the built-in mic of the iPad in order to get that low-fi effect that is typical of the original Beatboxing artists. To record our sample, we need to follow these steps (Figure 7.53):

1. Set the input source.
2. Set the "Threshold" level, which determines the level at which the sample recorder will start recording (set it to a level so the background noise will not trigger the recording).
3. If necessary, adjust the prerecording time. This will determine the amount of time that iMPC will record prior to the signal crossing the threshold. A longer time will help preserving the sharp transient of samples with a sharp attack.
4. Press the "Rec Arm" button and record your sample.

Before describing the editing process, we would like to give you some examples and ideas about Beatboxing sonorities. Because you will be generating the original samples with your voice, it is important to be able to identify the key features of each piece of the kit. Take a look at Table 7.5 for a description of the main characteristics of some of the most important sounds we will need to record.

Because one of the main sonic features of a Beatbox kit is the low-fi quality of the samples, we recommend using a good dynamic microphone. This will give you a fuller and "grungier" sonority. For extra low frequencies you can keep the microphone very close to your mouth, using the proximity effect as a sonic feature. Often we like to experiment with low-quality microphones (even the built-in microphone of the iPad!) to add extra distortion.

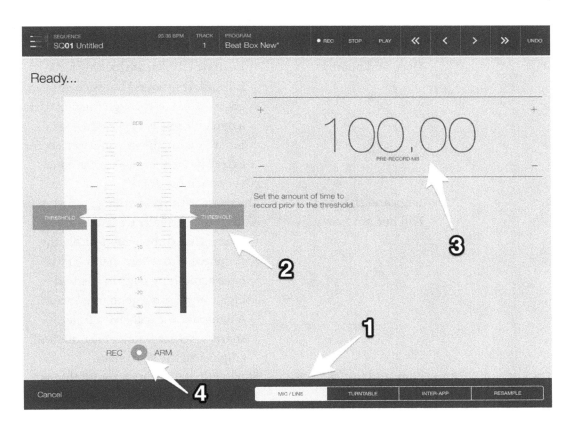

Figure 7.53
How to record a new sample in iMPC.

For some examples of Beatbox samples listen to Audio Examples 7.17–7.22 to compare a bass drum and a snare, respectively, recorded with the built-in iPad microphone, a dynamic microphone, and a high-quality condenser microphone.

Editing the Samples

After recording the samples, you need to make sure that the start and end points of the audio material captured are exactly what you had in mind. We followed the same procedure when we recorded the samples for the acoustic kit. Basically we need to trim the silence before and after the audio material we recorded. In the iMPC app the waveform editor opens up as soon as you press the Stop button during recording (Figure 7.54). To adjust the beginning and end of the sample use the Start (S) and End (E) tabs. As we learned earlier, make sure that the start is as close as possible to the beginning of the transient but not too close so that it could create a digital click. For the end of the sample leave enough space for the sample to decay naturally. To preview the sample tap on "Audition Trim" at the bottom of the screen, tap on "Done" when you are satisfied with the trimming you applied.

Once you have recorded the first sound, the process to record the other sounds is the same: Select a different pad and record the new sound.

Sample Parameters and Effects

Now that you have all your samples assigned to the pads, you can start working on the individual sounds. The iMPC offers good options in terms of sound control (Figure 7.55).

Table 7.5 Main Characteristics of a Beatbox Drum Kit

Piece of the Kit	Sonic Characteristics	Comments
Bass drum	• Full and edgy • Deep and boomy	The bass drum is a key element of a Beatbox kit. You are looking for a full and deep sound but with enough edginess to cut through the mix. Using a dynamic mic very close to your mouth (using the proximity effect for extra low frequencies) can produce very good results.
Snare	• Edgy and sharp • Full but not heavy	The snare is the piece that gives us more flexibility. Usually we like Beatbox snares that complement nicely the deepness of the bass drum. Try to have something edgy in the mid–high range with a good presence. A dynamic mic will give a fuller snare, whereas a condenser mic will emphasize the sharpness.
HH closed	• Edgy and focused	Start with generating a sound that is similar to a "zzzttttt" sound with your lips slightly apart and your tongue between your teeth. You can evolve from this by either experimenting with your voice or by using EQ during the programming phase.
HH open	• Sustained and full	Start with generating a sound that is similar to a "pssstt" sound with a longer sustain. Try to match the high-frequencies content with the HH closed sample so that they seem to belong to the same kit.
HH foot	• Fuller than the HH closed	This sample should be similar to the HH closed but a bit fuller and with a slightly longer sustain.
Crash cymbal	• Edgy and bright • Long sustain	This sound needs a long sustain, a bright but full tone, and a sharp attack.
Toms	• Full and big • Edgy with a bit of sustain	Electronic toms are probably the easiest to create with your voice. For anything more acoustic we recommend altering the sound during the programming phase with the addition of equalization and compression.
Noises	• Usually sharp attack and punchy short release	Typical sounds for these effects are the "T" and "K" sounds sequenced in rapid successions (drum'n bass style). You can experiment with several other noises that fit your style and patterns.

Sample-Based Synthesis

Figure 7.54
The Sample Edit page in iMPC.

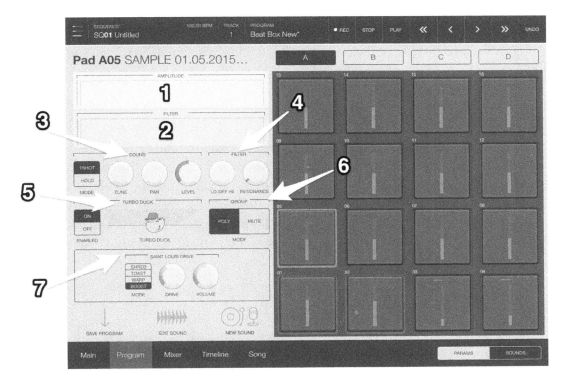

Figure 7.55
Effects and sample options in iMPC.

For each pad you can change the following characteristics:

1. The amplitude over time with attack and decay points.
2. The filter over time with attack and decay points.
3. Tune, pan, and volume.

Creating Sounds from Scratch

4. Filter and resonance. Moving the knob to the left will set the filter as a LPF, whereas moving it to the right will set it as a HPF.
5. Turbo Duck is a hard compressor/limiter
6. Group: Use this option to assign two or more pads to the same group in order to have only one pad of the group playing at the same time. This is useful for HH close/open effects.
7. Effects: The iMPC offers two effects, a drive/distortion and a ring modulator. Use the former to add punch and grunginess to the sample, whereas you can use the latter for a more creative approach and to shape further the pitch content of the samples.

As we learned, the Beatbox drum kit can vary and has an almost endless list of sounds, sonorities, and effects that can be created. It is important when you start though to keep your focus and have few points in mind. Here is a set of tips for a successful Beatbox kit:

1. Experiment with different microphones and mic-preamps. Remember that for this style not always the most expensive and high-end equipment will deliver the best results.
2. Start creating the basic sounds (bass drum, snare and HH) and then expand the palette with other pieces of the kit or with noise effects.
3. Use the filter to accentuate the low frequencies for the bass drum and the mid and highs for snare and HH.
4. Use compression to highlight the attack of the samples. For a more "live" effect use a hard limiter.
5. If you are not sure of the sounds you are looking for, leave the mic on and keep trying, then pick the best sound when editing the recording.

Creating Beatbox Patterns

Although creating a good kit is crucial, it is also important to create realistic Beatbox patterns that can take advantage of our sounds. When it comes to traditional Beatboxing there are a few styles that specifically relate to Beatboxing, such as Hip-Hop, Drum n' Bass, House Dance, and Techno. In Figures 7.56–7.59 you can find some examples of these styles. Listen to Audio Examples 7.23–7.26 to hear these patterns sequenced with the iMPC and an original Beatbox kit. Use them as a starting point but feel free to experiment and come up with your own variations.

Figure 7.56 Example of hip-hop groove for Beatbox kit.

Figure 7.57
Example of Drum n' Bass groove for Beatbox kit.

Figure 7.58
Example of House groove for Beatbox kit.

Figure 7.59
Example of Techno groove for Beatbox kit.

Working With Pitched and Sustained Samples

So far we have been working with percussive instruments that are not pitch based and that don't have audio material that needs to be looped. In this section we are going to learn how to create an instrument from an acoustic bass played with *arco*. A string instrument offers that ideal case study for practicing the creation of a more complex library, one that has to take into consideration the mapping of different pitches, looping of sustained notes, and sample switching based on up and down bow playing techniques. Let's start!

Recording the Samples

As we learned when working on the acoustic kit, the first step for creating a sample sound library is to record, in the studio, all the samples that we need for programming the instrument. In this case we have chosen an acoustic bass played with the bow. In particular, we are going to create a down and up bow sustained patch. As we learned earlier in this chapter, the planning stage for a successful sample library is crucial. Therefore, first you need to have a clear idea of which samples you will need to create the library. Look at Table 7.6 for a list of the samples we recorded for the acoustic bass library.

As you can see in Table 7.6, we have recorded pitches at intervals of half or whole steps. In general, it is recommended to record each individual note (every half step), but anything in the range of a whole step can be effectively interpolated by the sampler's engine. In the case of string instruments, make sure that you have samples of all the open strings, as they have a slightly different sonority compared with that of a pitch not played on an open string. It would be impossible for the sampler to correctly interpolate an open string sample from a not-open string sample. For this example, we have created only *mf* dynamics, but you can have a wide range of dynamics triggered by different velocities, if

Table 7.6 List of Samples for the Arco Bass Patch

Pitch	Down Bow Name	Up Bow Name	Comments
E1	Bass_Sus_E1_v1-Down -mf	Bass_Sus_E1_v1-Up -mf	For each Down and Up sample we also recorded two or three different variations that we have indicated with V1, V2, and V3.
F#1	Bass_Sus_F#1_v1-Down -mf	Bass_Sus_F#1_v1-Up -mf	
G#1	Bass_Sus_G#1_v1-Down -mf	Bass_Sus_G#1_v1-Up -mf	
A1	Bass_Sus_A1_v1-Down -mf	Bass_Sus_A1_v1-Up -mf	
B1	Bass_Sus_B1_v1-Down -mf	Bass_Sus_B1_v1-Up -mf	
C#2	Bass_Sus_C#2_v1-Down -mf	Bass_Sus_C#2_v1-Up -mf	
D2	Bass_Sus_D2_v1-Down -mf	Bass_Sus_D2_v1-Up -mf	
E2	Bass_Sus_E2_v1-Down -mf	Bass_Sus_E2_v1-Up -mf	
F#2	Bass_Sus_F#2_v1-Down -mf	Bass_Sus_F#2_v1-Up -mf	
G2	Bass_Sus_G2_v1-Down -mf	Bass_Sus_G2_v1-Up -mf	

needed, as we learned for the acoustic drum kit. The nomenclature system we have used for the bass is similar to the one we learned earlier in this chapter. Here's a breakdown of the name of the samples:

Instrument name: Bass
Articulation: Sus for sustained
Note and Octave: D2
Version: V1
Bowing direction: Down (or up)
Dynamic: *mf*

When recording pitched instruments, you have to pay particular attention to the tuning. For fixed-pitch instruments like the piano, the guitar, or the fretted bass, make sure that you regularly tune the instrument during the recording of the samples. For fretless instruments like the acoustic bass, you need to ask the performer to pay particular attention to the tuning of each note. Although small tuning problems can be easily fixed in postproduction, it is always better to get the best performance during the recording of the samples. In general, we recommend using a pitch corrector like the one in Logic Pro X (Figure 7.60) to make sure that the samples are in tune. Do not overcorrect the pitch

Figure 7.60
Sample pitch corrections in Logic Pro X.

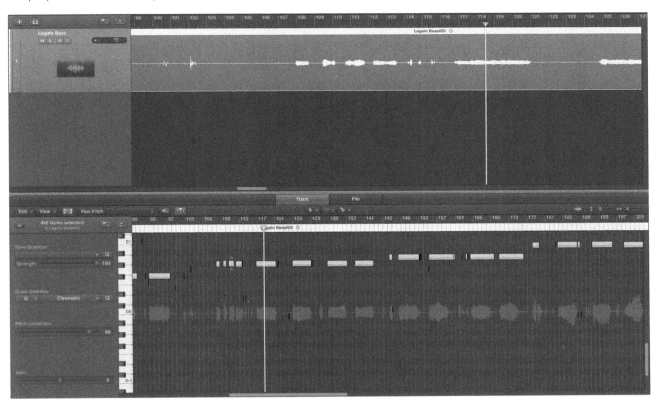

though, you want to have a natural sounding instrument; therefore just use a correction of 50% or lower as a starting point.

Another important factor to take into consideration is the consistency of the samples. For sustained instruments it is crucial to achieve a certain consistency among the samples so that you will have a smooth and natural-sounding switch among the notes triggered by your controller. Pay particular attention to the intensity, color, and attack/release of each note. Every sample in the same dynamic range and bow stroke must have the same characteristics. For this reason, we recommend always recording several variations to choose from during the programming stage. To make sure that you don't get confused with all the samples you will end up with, we recommend "announcing" before each sample its articulation, note and octave, version, and any other information that you might find useful during post production. Listen to Audio Example 7.27 for an excerpt from the recording session of some of the bass samples we did for this chapter.

Once the samples are recorded, go through them and pick the best ones. Again look for good tuning, cleanness, and consistency. Export all the selected samples using the suggested nomenclature system. Export them in the different folders organized by articulation, dynamic, and bowing direction. If necessary, before exporting them, use equalization and dynamic effects to obtain the desired sonic quality or to fix any problem that arose during the recording. In general, for acoustic instruments, we like to keep the sound as natural as possible and let the final user apply any additional effect.

Programming

With the samples ready to go, now it is the time to start programming our bass patch. The steps are fundamentally the same as those we learned for the acoustic kit, with a few important additions. First, we need to map our samples. To do so we recommend using the autoplacement tool in Kontakt. First drag all the samples into the mapping editor, then set up the rules according to your nomenclature system, and finally highlight all the samples and select *Automat Functions>Automat Selected* from the Edit menu. Make sure that each sample is assigned to the right key based on its pitch. Now we need to have the sampler interpolate all the missing pitches. In the case of the acoustic bass example we recorded the pitches listed in Figure 7.61. The gray keys are the interpolated pitches.

To have the sampler create the missing samples, we need to extend the upper limit of each region to cover the missing notes. To do so, simply drag the upper limit of the regions over the missing pitches. We recommend always extending the samples toward higher pitches because the interpolation will be more effective and realistic. This is because the sampling engine has to remove individual samples when extending a zone upward instead of having to add interpolated samples when extending it downward.

Looping

Once the samples are mapped to the right keys we need to open the waveform editor and not only make sure that each sample's start and end are perfectly set, but also program the loop points in order to have each note sustain indefinitely until the triggering key is released. Perfecting the looping technique usually requires a good amount of practice and experience. If you are new to this don't get discouraged; you will get better at it faster than you think. It is also important to point out that looping might not be always necessary even for pitched sustained instruments. This is the case for instruments like a piano, harpsichord, or guitar. For such instruments it is better to record the natural decay of the strings instead of trying to re-create their decays through a fading loop. The samples will

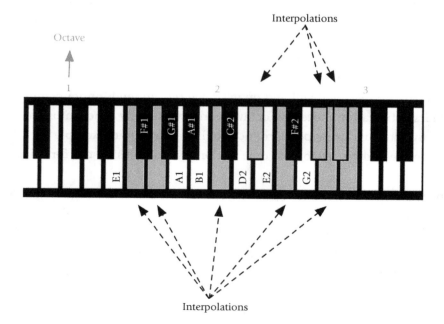

Figure 7.61 Original recorded pitches for sampled bass patch.

Sample-Based Synthesis

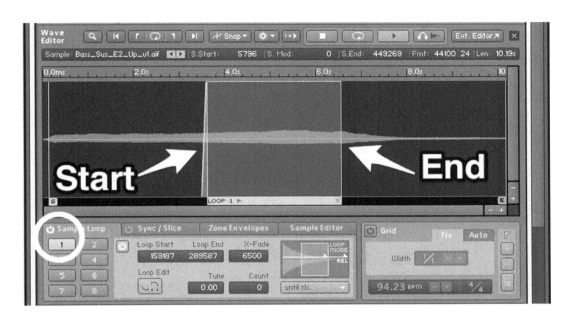

Figure 7.62
Editing a loop in Kontakt.

be longer, and therefore the size of the library will be bigger, but the results will be more natural. For string instruments, on the other hand, we need to loop the samples because, technically, a string instrument could keep sustained a note for a very long time.

To start looping our samples in Kontakt we need to open the Wave Editor. Select the first zone in the mapping editor and then open the Wave Editor in Kontakt. Here is where we can adjust the start and end of the samples. Kontakt offers up to a total of eight loop points that can be adjusted independently. To select the loop area click on the first loop option (number 1). The loop area is highlighted in yellow (Figure 7.62). The goal now is to define the loop Start and End points that give us the smoothest transition from the end back to the beginning of the loop.

Here are some recommendations to find the smoothest loop point:

1. Look for general areas where the amplitude of the End point matches the amplitude of the Start point. Big changes in amplitude are problematic for a smooth looping.
2. Choose the looping area in the second half of the sample. Usually starting around the half of the total length of the sample and ending around 3/4th is a good beginning.
3. Use the loop crossfade capability of your software sampler to help smooth the transition between the End and Start of the loop. In Kontakt you adjust the fade using the "X-Fade" value (Figure 7.63).

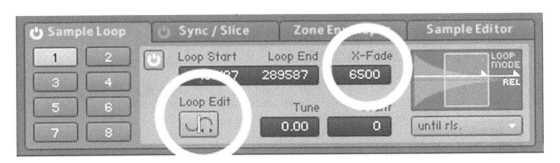

Figure 7.63
Loop crossfade.

Figure 7.64
Nonideal loop point.

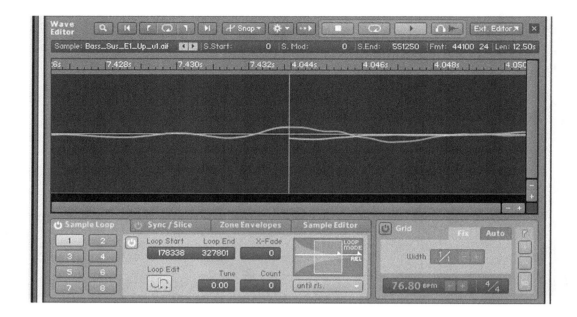

4. After finding an area where the Start and End work pretty well, zoom in and look for a seamless transition of the waveform between the two points. In Kontakt use the "Loop Edit" option (Figure 7.63) to easily find the perfect transition. In Figures 7.64 and 7.65 you can see a poor transition (the skip in the waveform will produce a digital click during the loop) and a good smooth transition (notice how the waveform is seamless), respectively.

Listen to Audio Examples 7.28 and 7.29 to compare the two loops shown in Figures 7.64 and 7.65.

Figure 7.65
Smooth loop point.

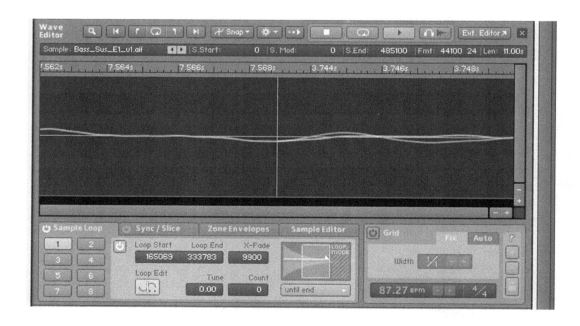

Figure 7.66
Envelope of the amplifier in Kontakt.

Programming

The programming part of creating a bowed acoustic bass involves mainly three tasks: setting the envelope of the amplifier (AHDSR), programming MIDI velocity to control volume, and assigning a MIDI controller to the Down/Up bow sample switch.

First let's set up the envelope of the amplifier. This task involves the addition of a modulator that controls the volume of the instrument. The modulator that we are going to use is an AHDSR. In Kontakt you can easily do so by right-clicking (or ctrl-clicking) on the volume knob in the amplifier section and select "Envelopes>ahdsr." A new modulator will be added at the bottom of the rack (Figure 7.66)

For an acoustic instrument we usually leave the attack, hold, decay, and sustain portions of the envelope as they are in order to keep the natural flow of the sound intact. We recommend, though, working on the release section, particularly for string instruments. If you have a short release, when you let the key go, the triggered zone will stop abruptly; this might result in an artificial effect that does not reflect how an acoustic instrument would sound. Try to have a natural release instead by slowly increasing the release time. For extra flexibility you can also assign the release parameter to a MIDI CC (Figure 7.67). This will allow you to dynamically control the release time depending on the phrase required by the passage you are performing. To do so in Kontakt, right-click (or ctrl-click) on the release knob in the amplifier modulation section (Figure 7.67) and select MIDI CC#. In the new section added at the bottom of the AHDSR, type in the MIDI CC number that you would like to assign to the release parameter. In this case we have used MIDI CC72 which is the standard one used to release control.

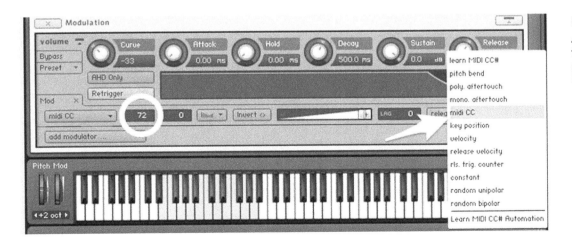

Figure 7.67
Assigning the Release parameter to a MIDI CC.

Figure 7.68

Grouping zones according to their bow articulation in Kontakt.

Once the amplifier envelope is set we need to make sure that our instrument is dynamic, which means that different MIDI velocities will correspond to changes in volume. To achieve this effect, we need to assign MIDI velocity to control the amplifier. The process is the same we learned for the AHDSR. Right-click on the volume knob in the amplifier section and select "External Sources>Velocity."

Finally, we need to be able to switch between the Down bow and Up bow articulations. To program this functionality, we are going to use the Group function we already used when working on the acoustic drum kit. The main idea is to group all the zones that contain Down bows in one group and then do the same for the zones containing the Up bow samples. In Kontakt select all the Down bow samples (we recommend switching to the List View in the Mapping Editor [Figure 7.68]) and by right-clicking on one of them select "Move zone(s) to new empty group."

Figure 7.69

Assigning the Down Bow group to start with a value range of 0–64 with MIDI CC20.

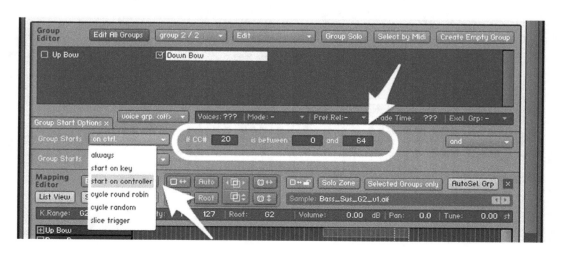

Sample-Based Synthesis

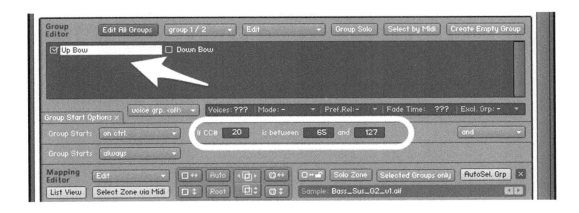

Figure 7.70
Assigning the Up Bow group to start with a value range of 65–127 with MIDI CC20.

Rename the new group "Down bow." Repeat the same steps for the Up bow zones. The technique that we are going to use is based on the idea that we can assign a MIDI CC to select the group we would like to activate and play. Therefore, by using, for example, CC20 we can choose to activate either the Down bow or the Up bow samples at any given time. To do so select the Down bow group, click on "Group Start Options" and under the Group Start tab select "Start on controller" (Figure 7.69).

Technically you can choose any MIDI CC to switch articulation, but we usually recommend using CCs that are undefined in the MIDI specification, such as the ones from 20–31. For this example, we are using CC20. Next set the range of values for the specific MIDI CC selected. Split the values according to the number of articulations you need. For example, if we have two articulations (Down and Up bow) we will use values from 0 to 64 for the Down bow articulation, and values from 65 to 127 for the Up bow articulations. If we need three articulations we will set the values as 0-42, 43-86, 87-127 and so on. Once the Down bow group is set you can proceed to program the Up bow group in the same manner (Figure 7.70).

Now, when performing with this instrument, if we want to play a Down bow sample we will simply bring CC20 below 64, and when we need to play an Up bow sample we will bring MIDI CC above 65.

Exercises

7.1 Using the samples provided on the website, create an acoustic drum kit with different levels of dynamics for each piece of the kit.

7.2 Create a Beatbox recording your own samples. Make sure to have at least the following sounds: bass drum, snare 1, snare 2, HH closed, HH open, Crash, Tom hi, Tom Low.

7.3 Using the Beatbox created in the previous exercise sequence four different drum grooves of eight bars each.

7.4 Using the samples provided on the website create an acoustic bass instrument with Down and Up bow articulations.

8

Physical Modeling

Introduction

The logical extension of additive synthesis is physical modeling synthesis, also called "component modeling," as they both approach the process of sound design through the assembly of a complex sound's individual parts. Additive synthesis uses many sine waves set to frequencies and intensities that create the desired harmonic/overtone series of a sound. Physical modeling is a departure from past techniques in that its sounds are based on mathematical models designed to create new sounds and emulate acoustic ones, often with impressive accuracy. The controls of a physical modeling instrument may look a bit unfamiliar: rather than offering a sound source followed by the usual envelope and filtering parameters we are accustomed to, a sound is divided into fragments of attack (e.g., mallet, pick, or breath) and resonance (e.g., beam, string or tube), each with parameters specific to their type. The approach is more about selecting a resonant object (resonator), causing it to vibrate with another object (generator) and using yet another object or objects (damper) to dampen or otherwise interrupt the vibration.

Analog Modeling

A first baby step into the modeling world came in the 1990s with digital instruments that used algorithms to emulate analog oscillators, envelopes, filters, and so forth. Although purists may still yearn for the real thing, instruments like the Nord Lead did a very respectable job of reproducing the sounds and modulation controls of their analog counterparts while eliminating issues like intonation instability common to analog oscillators. Later versions like the Nord Lead 4 (Figure 8.1) contain models that emulate standard

Figure 8.1
Nord Lead 4 Analog Modeling Physical Modeling Hardware Instrument (photo courtesy of Clavia DMI AB).

sine, triangle, pulse, and sawtooth waves plus waves that would be impossible to produce with an analog oscillator, made available in a wavetable-style menu.

As you may suspect, the mathematical calculations inherent to physical modeling are rather complex and thus demand a significant amount of processing power. When physical modeling was first introduced into commercial products in the 1990s, instruments were expensive and polyphony was limited because of processing already allocated to sound generation. Once personal computers were capable of efficiently managing the number crunching, polyphonic physical modeling software instruments became a viable part of our toolkit.

Physical Modeling in Hardware

One of the first mainstream hardware instruments to use physical modeling for its sound generation was Yamaha's VL-1 (Figure 8.2) released in 1994. At a time when sampling was considered the best method for realistic emulation of acoustic instruments, the VL-1 came along with sophisticated algorithms for patches that sounded as good as, if not better than, samples—especially with sounds with nonpercussive attacks that continue to challenge the limits of samplers today.

Although keyboard instruments are the most common MIDI controllers used today, with the release of the VL-1 it became apparent to Yamaha how data from a key strike limited articulation possibilities afforded by a technique like physical modeling. They developed the BC2 breath controller (also shown in Figure 8.2) as a way to circumvent this limitation. With a device like the BC2 in the keyboardist's mouth, articulation and velocity control can be made to sound more idiomatic of a wind instrument, leaving the keys solely for selecting pitch. It also allowed an unprecedented expressive control over any sound produced by the VL-1 that broke from the keyboard-centric paradigm.

Woodwind players would soon have a number of wind-controller options as this technology evolved, most notably instruments that look and feel like a clarinet or alto saxophone with a mouthpiece and keys arranged in a way that is familiar to them. A good modern example of this is the AKAI EWI-series Wind Controllers (Figure 8.3).

Physical Modeling in Software

There are many good examples of software physical modeling instruments, a few of which we explore in this chapter. The first we look at is Applied Acoustics Systems' (AAS) Tassman, which incorporates a variety of synthesis techniques including physical modeling.

Figure 8.2
Yamaha VL1 and the included BC2 breath controller (photo courtesy of Yamaha Corporation of America).

Figure 8.3
Akai EWI5000 Wind Controller (photo courtesy of Akai Professional).

Earlier in the chapter we described the stages of a physical modeling instrument: generator (attack), resonator, and damper. Figure 8.4 shows a simple patch designed in Tassman that uses a model of a percussion mallet (acting as the generator) exciting a beam (the resonator) in a manner analogous to that of a marimba. (Audio Example 8.1)

The modular nature of Tassman is such that designing an instrument follows an object-oriented programming model seen in the bottom portion of Figure 8.4. The performance user interface is on top. The signal path begins with a polyphonic velocity-sensitive keyboard input ("polyvkeys") that has three outputs: Gate on/off, Pitch, and Velocity (from top to bottom). In this example the Gate output triggers the mallet and the Pitch output regulates the pitch of the beam resonator. The various modulation inputs of both the generator (mallet) and resonator (beam) are shown on the right-hand side of Figure 8.4. These can be controlled from any number of modulator sources as well as the knobs on the user interface. For simplicity in this diagram, we used a "Constant" value of 1 (on a scale of 0 to 1) fed to the beam's input damper to allow maximum sustain when the "decay" control is turned all the way up.

Table 8.1 contains a list of the physical modeling generators in Tassman. Characteristics like the stiffness of a mallet or plectrum are controlled by user-definable inputted values. Logically one would pair a mallet with a beam and a plectrum (think

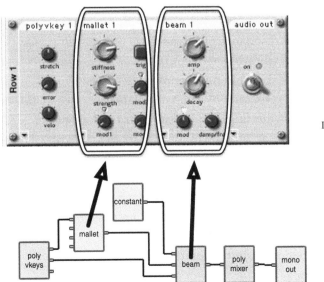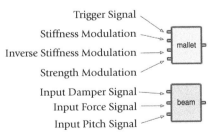

Figure 8.4
Basic Physical Modeling instrument designed in AAS's Tassman.

Table 8.1 Sampling of Generators in AAS Tassman

Generator	Description
Constant	Outputs a constant, user-inputted value
Mallet	Simulates a mallet striking an object; modulation includes stiffness of mallet head and strength of the impact
Noise mallet	This one takes a little imagination: it acts like a mallet, but the impact is both the mallet characteristics and white noise
Plectrum	Simulates the plucking of a string with a finger or a pick; modulation includes stiffness of the plectrum and strength of the pluck.

guitar pick) with a string, but mixing these up is where physical modeling can get interesting.

To actually result in a sound, these generators must be paired with a resonator (see Table 8.2). Each resonator has user-definable and modulation-controllable characteristics to customize its timbre. Names like "Bowed Beam" and "Bowed Marimba" suggest unconventional generator–resonator combinations that can produce inspiring new sounds.

Table 8.2 Sampling of Resonators in AAS Tassman

Resonator	Description
Beam	Simulates the resonance of a metallic beam of a user-definable length (e.g., glockenspiel, vibraphone, etc.) vibrating with control over fundamental frequency, decay, and the number of modes present
Marimba	Simulates the resonance of a wooden beam of a user-definable length vibrating (e.g., marimba) with control over fundamental frequency, decay, and the number of modes present
Membrane	Simulates the resonance of a drumhead of user-definable dimensions vibrating with control over fundamental frequency, decay, and number of modes present
Plate	Simulates the resonance of a metallic plate with user-definable dimensions vibrating with control over fundamental frequency, decay, and the number of modes present
String	Simulates the resonance of a string of a user-definable length with control over fundamental frequency, decay, and number of modes present. The user can also define the location where the string is plucked and the location of the pickup.
Bowed beam, marimba, membrane, and string	Same as previously, but tailored for using a bow as a generator

Platypus of Sound?

Huh? A platypus? As you may know, the platypus is one of the strangest animals in existence: It is a mammal, but it lays eggs; it has a duck's bill, a beaver's tail, and feet like an otter! Okay, but what does this have to do with sound? Physical modeling makes it possible to do something analogous with synthesizers. How about a guitar that is played with a mallet and resonates like a cello? And while we are at it, let's give it wooden strings . . . or strings that are part wood and part glass? Yes, we are getting toward the end of the book but we assure you, our sanity is still intact (mostly). With the acoustic segments of an instrument reduced to algorithms, assembling them into unnatural and physically impossible combinations is simple and produces sounds that are at once organic yet novel.

Logic's Sculpture as a Case Study

The seemingly impossible scenarios just described become reasonable because we have mathematical models that describe the vibration and damping characteristics of materials rather than the real thing. Figure 8.5 is a screenshot of Logic's string modeling physical modeling instrument called Sculpture. As an overview, the interface is divided into three columns: The left-third deals with exciting and damping the virtual string, the center-third is where the material the string is made from is defined, the top portion of the right-third has familiar modulation controls seen on other instruments, and the bottom-right is where the resonating body characteristics are defined. So the real physical modeling components are (1) where and how the string is disturbed and damped, (2) the kind of material is the string made of, (3) the familiar synth modulation controls, and (4) the size and material of the resonant body on which the strings are mounted (see Table 8.3, Audio Examples 8.2 and 8.3).

The approach to and results from working with physical modeling are completely different than the techniques we have encountered so far. Let's now transition to the producer point of view and look at creative ways of incorporating physical modeling into your creative synthesis.

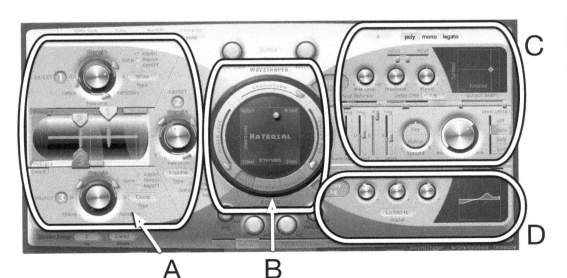

Figure 8.5
Logic's Sculpture Physical Modeling instrument.

Table 8.3 Breakdown of Sculpture's physical modeling Controls

Section	Description
A	Object 1 contains a list of methods for stimulating the string: impulse, strike, pick, bow, noise, and blow.
	Objects 2 and 3 have the same lists as object 1 with a few additions for stimulating, disrupting, or damping: disturb (damping), bouncing, bound (boundary that limits string vibration), mass, damp, and external (input).
	In between the three object control knobs (which are used to customize characteristics of the objects) is a graphic of a string with sliders labeled 1, 2, and 3. The sliders allow you to place the corresponding object along the length of the string, The unnumbered rectangular sliders at the top and bottom of the string "window" represent guitar-style pickups that capture harmonic content relative to placement; closer to the ends of the string will favor upper partials, toward the center will favor the fundamental.
B	Material composition of the string; an *X/Y* matrix that allows morphing of the string's material smoothly from nylon to wood to glass and to steel. The *Y* axis (nylon/wood to steel/glass) affects inner loss, and the *X* axis (nylon/steel to wood/glass) affects the string's stiffness.
C	Standard synth controls: delay, envelope, level. . . .
D	Body resonance: guitars, dobro, banjo, mandolin, ukulele, kalimba, double bass, cello, violin, flutes. . . .

The Producer Point of View

We consider physical modeling to be one of the most exciting and inspiring synthesis techniques. The idea behind it is really fascinating, an approach to synthesis that, we believe, is going to play a major role in the evolution of new sonorities and styles of music. Because of its "deconstructive" approach in regard to how sounds are generated, amplified, and ultimately perceived, physical modeling gives us a lot of freedom in creating some really original and interesting sonorities. Even though it can be used effectively to re-create realistic acoustic sounds, it is the more creative part of physical modeling that we find particularly captivating and appealing. In the next sections we discuss the main software synthesizers we use and recommend and also the techniques involved in how to create original and inspiring patches with this great method of synthesis. Then we will learn some compositional techniques and musical styles for which physical modeling patches are particularly indicated. Let's start!

Some of Our Favorite Physical Modeling Synthesizers

As we learned in the previous chapters, hardware synthesizers have usually led the pack when it came to experimenting with new forms of synthesis. The most popular hardware physical modeling synthesizers were the Yamaha VL series (VL1M, VL7, and VL70-m). These machines were (and still are) the perfect companion for Yamaha's wind controllers. The higher flexibility offered by physical modeling in terms of real-time performance control means that woodwind players, using the Yamaha WX MIDI controllers (or the

AKAI EWI series) can translate almost every nuance from the acoustic world to the digital domain.

Although this has been the case for physical modeling to a certain degree, we feel that it is really with the "democratization" of powerful personal computers and software synthesizers that this type of synthesis has been able to become mainstream for contemporary producers around the world. With more powerful machines, physical modeling has become more accessible, inspiring writers and producers around the world. Some of our favorite software synthesizers are the ones that enhance our creativity, allowing us to create extremely original sonorities that are otherwise not easily obtainable with other types of synthesis. In the last part of this chapter we are going to focus on letting your creativity run free using physical modeling. Although we really like the potential that this type of synthesis has in re-creating acoustic sonorities, we find that, for a contemporary producer, physical modeling's most fascinating aspect resides in its ability to create what we call "Franken-sounds," meaning sounds that borrow parts, techniques, and textures from the acoustic realm of sound to combine them in a sort of genetically modified mix. For this sort of work we have some tools that we find particularly useful such as Studio Strings VS-2 (AAS), Chromaphone (AAS), and Sculpture (Logic Pro X). Although there are several others available, as producers and writers we always like to strive for the right balance between powerful features and usability. We like to use tools that allow us to reach the maximum creativity and originality in the minimum amount of time. These three software synthesizers offer an excellent set of features with an easy and straightforward user interface. In the next sections of this chapter we will learn how to create and manipulate some typical physical modeling patches and how to create more original variations.

Main Typologies of Physical Modeling Sounds

Even though it can be used to create endless categories of sounds, there are a few that we find particularly interesting and that are more common for contemporary production styles. In Table 8.4 you can see a brief description of their sonic characteristics and usage:

Sound Design and Production Techniques for PM Patches

Your goal in the next sections is to learn how to create effective and useful patches for your next productions by using physical modeling and at the same time familiarizing yourself with some of the most common production and compositional techniques for which physical modeling can come in particularly handy. As our guideline we will use the list of Table 8.4. Keep in mind that these are some of our favorite options but, as you should know by now after reading this book, the possibilities are almost endless and the boundaries are there for you to explore and break! For these examples we use three main software synthesizers: Logic Pro's Sculpture, AAS's Chromaphone, and String Studio VS-2 by AAS.

Percussive Pluck Patches

The percussive and pluck patches are among our favorites when it comes to physical modeling. They can be used in a variety of styles ranging from electronic pop to acoustic and

Table 8.4 Typical PM Patch Categories

Category	Sonic features	Compositional and production features
Percussive plucked	The driver is usually a plucking tool (softer or harder depending on the needs), or a mallet/stick; the resonator could be a string or a membrane. The damper can be usually set to frets or hands. The overall tone is cut through with an edgy sonic print that can be made softer by working on the positioning of the driver.	• These types of patches can be very effective for rhythmic parts that need to also outline harmonic or melodic passages. It is particularly indicated for repetitive steady rhythmic passages that have the double role of moving the groove forward and at the same time supporting the harmonic motion.
Pads (bowed and blowed)	The driver can be based on bow or blown air. The resonator can be a reed, string, or open tube.	• These pads provide a wide range of colors. They can be smooth and airy, providing a light and suspended feel, or they can be thicker and more aggressive, grounding the sound more. • They usually work well when played in octaves, or open fifth or fourth, enhancing the openness of the musical situation.
Percussive pads	The driver can be a picking tool, and the resonator can be set to reed, tube, or strings. They are a combination of the previous two categories.	• They provide a nice balance between a sonority with a percussive attack that cuts through the other instruments while, at the same time, providing the necessary sustain to better support the harmonic motion. • Use these patches to provide the backbone of a piece.
Leads (bowed and blowed)	The driver can be a bow or blown air. The resonator can be a reed, tube, or strings. Although similar in nature to the pads category, the leads are usually more "esoteric" and creative.	• Leads can be a very creative category. They can range from lighter and airy sonorities and stretch all the way to more edgy ones. • Effects like reverb and delay help enhance their original spaciousness, and saturation and distortion can bring out their edginess and roughness.

Table 8.4 Continued

Category	Sonic features	Compositional and production features
Ambient/cinematic	This is the most creative category, in which sustained pads are combined with rhythmic evolving (or sequenced) more percussive elements. The driver can be anything (use harder material for more percussive effects). Use membranes or plates for the sustain parts to get more eerie effects. LFOs can be assigned to control both the driver and the resonator, creating ever-changing variations.	• These are patches that can express the full potential of physical modeling. By combining rhythmic elements with sustained components that change over time, you can create very interesting sonorities. Usually they tend to be fairly thick and majestic, but you can also obtain lighter and smoother effects by using bowing drivers and strings, or membranes as resonators. • The addition of effects such as modulation, delay, reverb, and distortion/saturation are essential in order to exploit the full potential of this category.

even folk. To create this patch, we are going to use Logic Pro's Sculpture. First, we want to make sure that we start from a default patch. Then we turn off Voices 2 and 3 so that we can concentrate on Voice 1 (V1) parameters. For these types of patches, we like to set the V1 exciter type to either gravity strike or strike, even though pick would also work. The latter will give you a slightly mellower sound whereas the two strike options will have a more energetic attack. Now we change the relative positions of the pickup and the driver to fine-tune the overall color of the signal produced. Generally moving the driver away from the pickup will result in thinner sounds. For this patch we use a setting in which the driver is positioned at the edge of the pickup (Figure 8.6).

Now we can start working on the string parameters. The one parameter that can affect the overall sonority of a Voice is the material (Figure 8.7). The material pad allows us to morph freely from one material to the other with any degrees of variation among nylon, wood, steel, and glass. For this particular patch we have chosen a position halfway between nylon (which gives a warmer sonority) and steel (which provides an edgier and slightly aggressive touch).

Feel free to experiment with the material, as it is one of the parameters that can have a bigger impact on the final sonic characteristic of your patch. Before moving to the other parameters, we recommend experimenting with at least one other object. We particularly like to have another object interfering with the vibration of the string by using a "Disturb2" setting (Figure 8.8). Be careful not to overuse the disturbance because it could create some interesting but not always wanted effects. In this case we are positioning the interfering object away from the center of the strings, and we also use a low strength in order to have a milder effect. Notice how the addition of the object creates more harmonics and brings the complexity of the patch to a higher level.

Creating Sounds from Scratch

Figure 8.6
Positioning the driver (Gravity Strike) and the pickup in Sculpture.

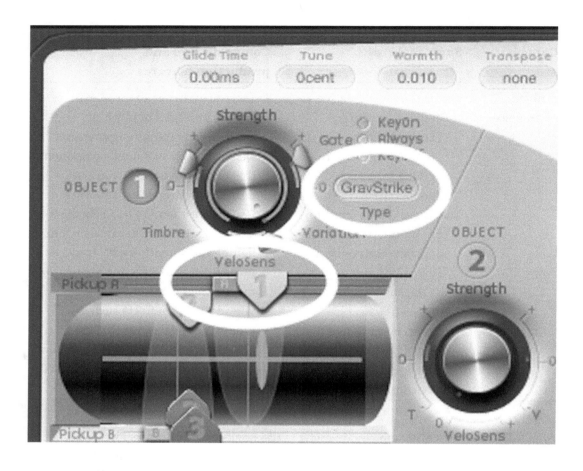

In Sculpture you have access also to a third object that can be used as a driver or damper. Once the basic sonic characteristics of the patch are in place, you can work on the envelope of the amplifier. Here we have control over the usual ADSR section. For this type of patch make sure that you have a very short attack. The release can be either very short (giving a more *pizzicato* effect) or a bit longer (around 160 ms) to give a bit more body to the sound. We like to control the attack time with MIDI velocity. In this case we have assigned velocity to the attack portion of the amp envelope (Figure 8.9).

Figure 8.7
Choosing the right material.

To add more "byte" to the sound we recommend using the Waveshaper section. This unit allows you to impose nonlinear curves to the signal coming from the pickups and amp envelope. You can choose from a smother effect (Soft Saturation) all the way to "Scream"! The end result is that, with the addition of the Waveshaper, you can add digital saturation or distortion. To improve even further our patch try adding a bit of delay. For percussive patches a delay can really bring your patch to life by adding more character and personality. For this patch we are using a delay synchronized to the tempo of the host DAW, with a rhythmic subdivision set to quarter-note triplets. In Sculpture we like to use the spread and groove parameters to achieve a wider effect on the stereo image and, at the same time, a slightly more interesting rhythmic effect (Figure 8.10).

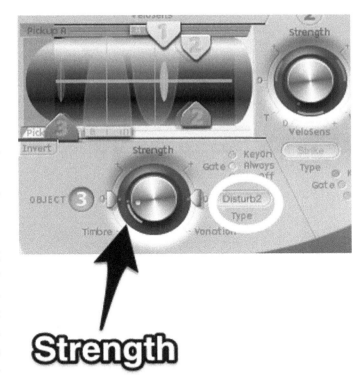

Figure 8.8
Inserting a disturbing object.

As we know by now, equalization always plays an important role in shaping the character of a sound, and it is one of the fundamental tools at our disposal when looking for originality. Because part of the charm of physical modeling is the ability to apply models of different instruments to our original patches, Sculpture offers the possibility of applying the equalization of different instrument bodies to the signal coming out from the three objects and the amplifier envelope. In this case we are going to use the body acoustic response of an acoustic guitar (Figure 8.11).

Feel free to experiment, as these settings can have a really big impact on the sonic characteristics of the patch. Listen to Audio Examples 8.4–8.6 to compare the same patch with the following Body Eq. applied: bass flute, ukulele, and cello.

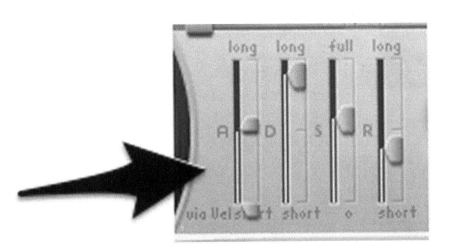

Figure 8.9
Assigning MIDI velocity to control the attack time.

Creating Sounds from Scratch

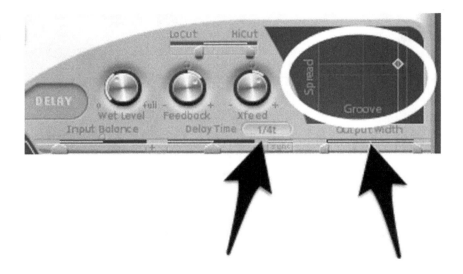

Figure 8.10
Adding a delay to our patch.

At this point we have a good percussive patch that can serve well in a variety of situations. To add a bit more variation we recommend using the Velocity Target pane to control different parameters with the MIDI velocity of a note. For this patch we set Target to Object 1 Strength and Target 2 to Waveshaper Variation (Figure 8.12).

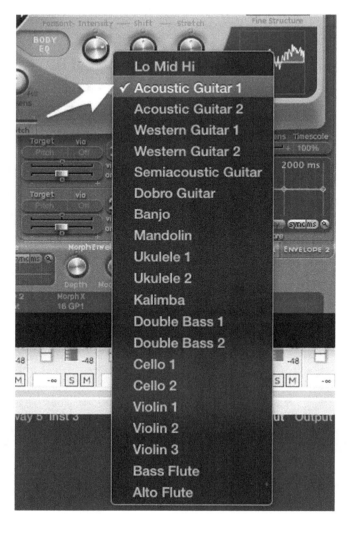

Figure 8.11
The Body EQ section in Sculpture.

Compositional Considerations

Physical modeling percussive plucked patches are usually effective when used in arpeggio phrases with a pedal note in the bass that can move against a repetitive chord. Look at Figures 8.13 and 8.14 and listen to Audio Examples 8.7 and 8.8 for two samples of this technique. In both examples we have used the patch we constructed in the previous paragraphs.

Another pattern that can work if this type of patch is used is the one shown in Figure 8.15, where the syncopated *ostinato* creates a nice contrast with the 1/4 delay.

Of course, don't forget that adding additional effects to the patch can open up a whole new area of fun creativity. For example, adding some pedal simulators and analog delays can drastically change the original percussive patch. Listen to Audio Examples 8.9 and 8.10 to compare the pattern shown in Figure 8.15 sequenced with a plain patch first and with the addition of a series of guitar pedal simulators.

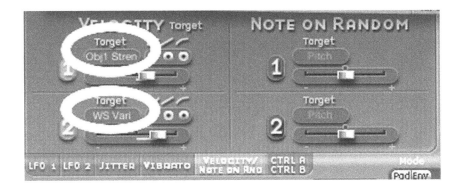

Figure 8.12
Controlling variations with MIDI velocity.

Pads

Pads are another category of sounds for which physical modeling can be very effective and inspiring. This type of synthesis can create a fairly wide set of string, pipes, and noise-based pads that can fit well any type of electronic or contemporary pop production. Software synthesizers such as Sculpture and String Studio VS-2 can be excellent tools when you are creating all sorts of pads. They both have an excellent choice of drivers, resonators, and controllable parameters. For this example, we use String Studio (SSt) because it offers an easy and intuitive interface without sacrificing the sonic capabilities. To start creating a sound in SSt we need to enter the edit mode by selecting Edit from the three options at the top of the main window, as shown in Figure 8.16.

Figure 8.13
Example of rhythmic pattern for a percussive plucked patch.

Creating Sounds from Scratch

Figure 8.14
Another example of rhythmic patter for a percussive plucked patch.

From the Edit menu we control and modify the main components of our virtual instrument. First, we are going to work on the exciter (Figure 8.17). For a pad patch we can use either a bow or a bouncing mallet. Although they are both suitable for this type of sonority we like to use the latter because it adds more variation and character to the patch.

By changing the mass and the stiffness of the exciter we can change the sonic characteristics of the patch. A higher mass will result in a deeper and "heavier" sound, whereas a lower mass will result in a more ethereal sound. On the other hand, higher values of stiffness will add a more metallic and high-pitched character, and lower values will give you a more subdued sonic character. Use the velocity parameter to control the intensity of the bounces. For this patch we decided to use medium settings across the board in order to have enough of the bounces present but at the same time preserving the necessary smoothness for the pad patch. All three parameters can be controlled via keyboard

Figure 8.15
A third example of rhythmic patter for a percussive plucked patch.

Physical Modeling

Figure 8.16
String Studio main menus.

positioning (Key) and velocity (Vel). Use the damping knob to reduce the extent of the vibration of the resonator. For a pad type of patch we suggest leaving the maximum setting.

The positioning of the exciter and the damper can be controlled from the geometry panel (Figure 8.18).

The position of the exciter has an effect on the frequency response of the patch. In this case, moving it more toward the right increases the low frequencies. In the same way, changing the positioning of the damper has an effect on the sustain of the sound. For a pad-like patch we usually turn it toward the left because we need a full sustain. Try to turn the pickup option "On." This will give a slightly more "electric" tone to the patch. Experiment with the position of the pickup. In general, moving the knob more toward the right will translate in a slightly thicker patch.

Strings and damper have their own parameters that can be set from their respective panels (Figure 8.19).

For a pad-like patch you need to set up the decay and damping parameters of the string all the way to the right. This allows the string to keep vibrating and giving a longer sustain. The damping parameter controls the material of the string by changing the amount of high frequencies generated by the

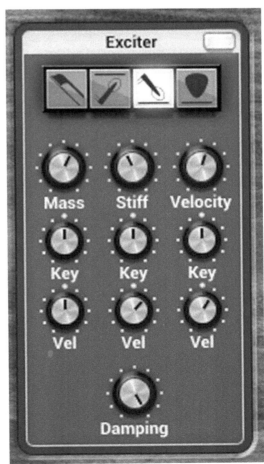

Figure 8.17
The Exciter settings in String Studio.

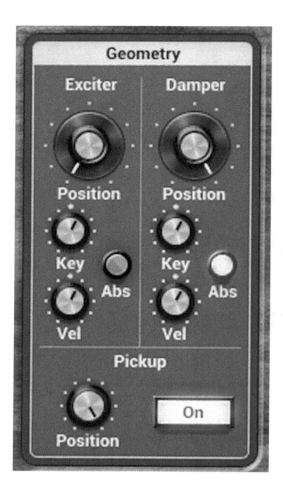

Figure 8.18
The Geometry settings in String Studio.

vibration. If you move it counterclockwise the sound will get darker and eventually extremely muffled. Use the release knob to control the vibration after you release the key of the controller. This behaves in a similar way as the release stage of a more traditional ADSR envelope. Use the level knob to control the overall volume of the string. The Inharm parameter allows us to control the inharmonic behavior of the string. In its leftmost position the string will vibrate in a perfectly harmonic way; turning the Inharm knob clockwise will increasingly detune the partials toward higher frequencies, resulting in a slightly detuned pitch.

The damper module allows us to control the parameters of the rapid attenuation of the string (such as the felts of a piano or the fingers of the guitarist). The mass and stiff knobs adjust how the damper interacts with and absorbs the energy of the vibrating string. A higher mass will stop the vibration in a shorter amount of time. Higher settings for the

Figure 8.19
Strings and Damper settings.

stiffness will result in a more metallic and vibrant tone. The damping knob at the bottom of the control panel is used to change the overall damping applied to the string.

From a physical modeling point of view it is particularly interesting to analyze how the body control panel (Figure 8.20) can affect the sound of our pad. As you can see, there are four body types available: piano, guitar, violin, and drum. Remember that the type and shape of the body can have a big impact on the final sonic characteristics of our patch. For this patch we picked a guitar-shaped body because its large sound hole allows us to add a bit more low frequencies. For each body type you can also select its size: tiny (T), small (S), medium (M), large (L), and huge (H). Smaller sizes result in a shift toward higher frequencies. Use the decay and the two filters (low cut and high cut) to control the material of the body. For this example, we chose a medium decay setting that slightly emphasizes the mid–low frequencies. We used the mix knob to balance the original signal (left position) from the vibrating string and the signal filtered through the body module (right position).

Figure 8.20
The parameters that control the body of the instrument.

To achieve the best results in creating a pad patch, we always like to use some audio effects in order to add a more cohesive ambient and some extra colors. Although the physical modeling pad we designed so far has a good personality we are going to add some effects from the string studio rack (Figure 8.21).

For a pad we recommend using modulation effects like chorus and phaser to blend and render more mellow the raw output of the physical modeling synth. In this case we have added a vintage chorus and a phaser. We also used a Hall reverb to place the patch in a more "spacey" environment and to add a bit more sustain to the patch. Finally a simple three-band EQ serves the purpose of boosting a bit the high frequencies and cut a little bit around 100 and 850 Hz for additional clarity.

Listen to Audio Examples 8.11 and 8.12 to compare the physical modeling pads with and without the addition of the effects.

Ambient and Cinematic

The category of ambient and cinematic sounds is one of the most fun to program and design. Here we can have our imagination run free, and physical modeling is one of the best tools to tackle this task. To create one of these patches we use a slightly different approach. Conceptually we first dissect the main patch in two different elements, then we

Figure 8.21
The String Studio effect rack.

create the two components in two different physical modeling synthesizers, and finally we layer the two components in one single combined patch. This technique allows us to use the most appropriate synthesizer, taking full advantage of each engine's best features. For this particular patch we decided to create two subpatches with very different, and yet complementary, sonic characteristics, as shown in Table 8.5.

As you can see, the two subpatches have very different characteristics. The layering technique is extremely useful when trying to create complex patches that need to evolve over time. Let's start with subpatch 1. For this one we elected to use Sculpture, mainly for its great morphing and modulation features.

The patch settings can be seen in Figure 8.22.

The main generator section is based on three sources: source 1 is set to "bow," source 2 is set to "blow," and the third one is set as a damper with the "bound" option. The first two sources allow us to create a pad-like patch, with some interesting sustained

Table 8.5 Ambient/Cinematic Subpatches Descriptions

Ambient/cinematic	**Sonic characteristics**	**Synthesizer used**
Subpatch 1	Evolving	Logic Sculpture
	Sustained	
	Spacey	
	Morphing	
	Pad-like	
Subpatch 2	Rhythmic	String Studio VS-2
	Percussive	
	Setting the beat	
	Energetic	

Physical Modeling

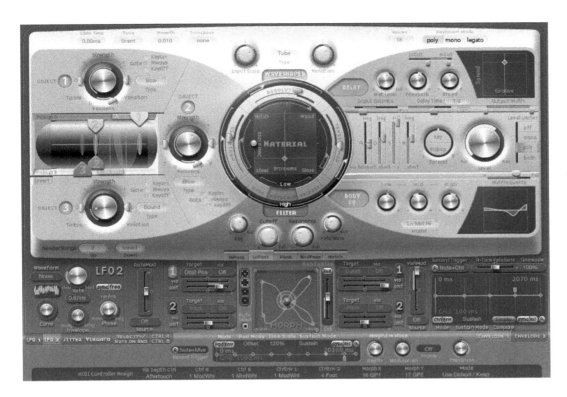

Figure 8.22
The first layer for a cinematic patch in Sculpture.

characteristics. The third one serves as the disturbance that gives us control over some of the more peculiar sonic features of the patch. The main goal and feature of this is to provide a constant change in motion to the combined sound. Therefore we need to take full advantage of Sculpure's morphing capabilities. One of the best features of this synthesizer is the ability to program a set of parameters that can be morphed over time. The parameters that can be morphed are color coded in orange, as indicated in Figure 8.23 by the white arrows.

Among the most important parameters that can be morphed over time, we suggest using the following:

- Strength of the source
- Position of the source
- Cutoff and resonance of the filter
- Tension modulation
- Media loss

To modulate the aforementioned parameters, we are going to use the morph module located in the lower part of the Sculpture window (Figure 8.24).

To morph between different settings, first we need to create snapshots of our patch with different variations for the aforementioned parameters. As you can see in Figure 8.24, there are five locations that can hold different variations of our patch. The one in the center is usually reserved for the starting point of our morphing path. In each corner of the square there are four more locations that can have different settings for each of the morphable parameters. The first step is to create the main sonic landscape that will be used in the center position in the morphing module. Once we are happy with

Creating Sounds from Scratch

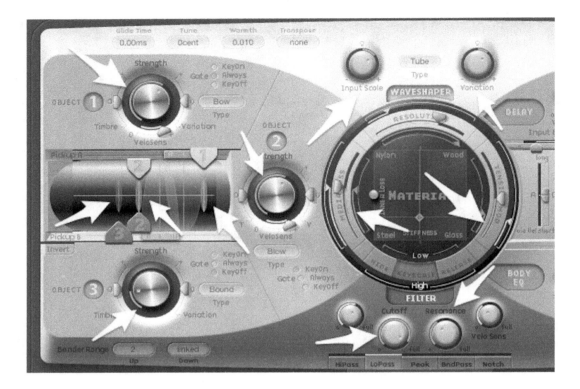

Figure 8.23
Morphable parameters in Sculpture.

the parameter settings for the center position, we recommend copying these settings for each of the four additional locations named A, B, C, and D. Make sure to click on the center location and then right-click on the morphing module and select "Copy selected Point" (Figure 8.25).

Once you have copied the main parameters of your patch, you need to paste the settings to the other locations in the morph module. To do so right-click on the module and select "Paste to all Points." Now all the four additional points, A, B, C, and D, will have the same parameters settings as the main location in the center. At this point you can manually change each location parameter to create four additional variations that then you will use to alter your final patch over time. If you don't feel like changing all the parameters manually, you can use the very useful randomization option as seen in Figure 8.26.

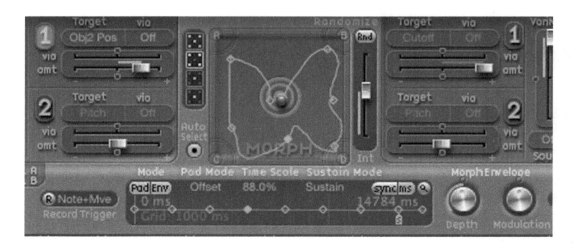

Figure 8.24
The Morph module in Sculpture.

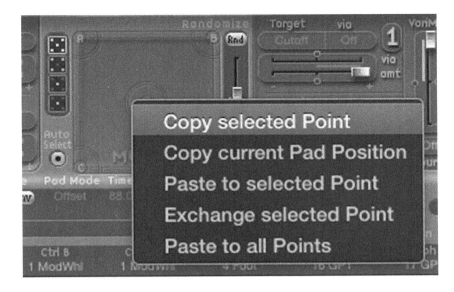

Figure 8.25
Copying the main parameters to the alternate locations in the Morph module in Sculpture.

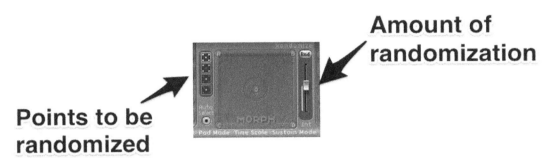

Figure 8.26
Randomization of the main parameters in the Morph module in Sculpture.

To use the randomization option, first you need to select the points that you want to randomize on the left-hand side of the morph panel. You have the option of randomizing the center only, the four corner points only, or all five points. In general, we recommend selecting the four corner points. After choosing the points, you need to select the amount of randomization you want for each of the points you selected. You do so by using the slider on the right-hand side of the morph panel. Try starting with some conservative settings first, anywhere between 25% and 40%. To create the randomization, simply tap on the random button ("Rnd") on top of the slider. Spend some time fine-tuning each location according to your needs. You can do so by either continuously pressing the randomization button or by selecting each location and manually changing the parameters. For this patch we created four variations using the randomization technique first and then manually changing the parameters of each location.

Listen to Audio Examples 8.13–8.17 to compare the main sound of the center location first and then the four variations.

Feel free to experiment. Each variation should be different enough from the center location but not completely different in order to maintain a certain level of consistency in the patch. If you are satisfied with your variations, it is time to program how the patch is going to morph between the different locations. To do so you can either manually move the morphing envelope points in the morph module or manually draw connecting lines

Creating Sounds from Scratch

Figure 8.27
Creating the Morph path.

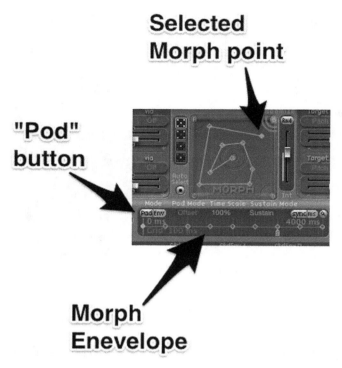

between the different locations. If you choose the former method, first make sure that you disable the "pod" button, as shown in Figure 8.27, then manually move each point of the morph envelope in any location that you would like your patch to morph it over time.

To record the morph path manually you need to make sure that the "pod" button is selected, then select the recording option that best suits your style by right-clicking next to the record button, as shown in Figure 8.28.

Now press the record enable button, play a key on your MIDI controller, and start moving freely the red "pod" button inside the morph module area with your mouse.

When done, deselect the record button. The new morphing path will show inside the morph module area (Figure 8.29).

Open to Audio Example 8.18 to listen to the patch with the morphing path shown in Figure 8.29.

Working on the Second Layer

As mentioned earlier, a typical cinematic/ambience patch is composed of two or more layers. Although in this case the first layer provides the foundation and the main connecting sonic material, the second layer is going to be dedicated to carrying the motion and the rhythmic aspects of the full patch. For this example, we chose String Studio VS-2 for two main reasons: its capability of creating short and percussive physical modeling

Figure 8.28
The record button and the record options for the Morph module.

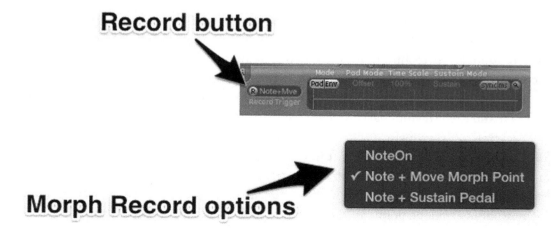

sonorities, and its flexible and powerful integrated rhythmic sequencer. Before exploring in more detail how we can use a built-in sequencer to create rhythm patches and layers, let's take a look at how to construct a more percussive patch by using the physical modeling in String Studio.

In Figures 8.30 through 8.32 you can see all the different settings for this second layer patch. Specifically notice how we set the exciter to a pluck (Figure 8.30). This allows us to

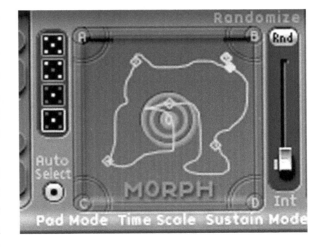

Figure 8.29
A new morphing path recorded using the mouse.

have a more percussive type of sonority. We also set the stiffness to a fairly high value. To give less importance to the pitch characteristics of the sound, we used a LPF with a cutoff frequency of around 2.5 kHz. For the body we picked a medium violin shape that allows us to add an edgier and cutting-through effect. A little bit of vintage distortion can accent the edginess of the rhythmic percussive pattern.

The arpeggiator in String Studio (Figure 8.33) allows us to set sixteen steps based on a predetermined rhythmic subdivision. The subdivisions can range between a quarter note all the way to a thirty-second note. The rate (speed) of the arpeggio can be set freely or synced to the tempo of the host DAW. In general we recommend using the sync option because it will make it easier to sequence parts that fit the groove of your tracks. For this patch we chose to use eighth notes as the basic rhythmic subdivision.

Figure 8.30
Settings for the second layer of the cinematic/ambience patch in String Studio: Exciter, Damper, LFO, and Body.

Figure 8.31
Settings for the second layer of the cinematic/ambience patch in String Studio: Geometry, String, Filter, and Distortion.

Figure 8.32
Settings for the second layer of the cinematic/ambience patch in String Studio: Termination, and Envelope.

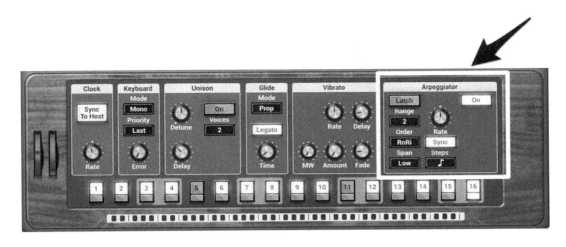

Figure 8.33
The "Play" screen with the Arpeggiator on the right in String Studio.

Use the range and span options to select the range over which the played notes are transposed and the direction of the transposition, respectively. Use "Low" for downward transposition, "High" for upward, and "Wide" for transposing upward and downward. The arpeggio pattern is determined by the "Order" parameter. Experiment with your favorite settings; you will be amazed by the many combinations and how much pattern variety you can create with an arpeggiator. Finally, you need to "turn on" the steps you would like to have played by the arpeggiator. As a default you have all the sixteen steps activated, but we usually like to mix things up a little bit by skipping a few steps in odd places. In this case, for example, we are not using steps 5 and 11. Listen to Audio Example 8.19 to listen to the second layer patch by itself.

Now that we have the two layers ready to go, we need to assemble them in one single patch. Depending on your DAW of choice, you will have different options. One of our favorite methods is to create a "stack track" in Logic Pro (Figure 8.34).

Figure 8.34
The final track stack in Logic Pro with the two layers we created.

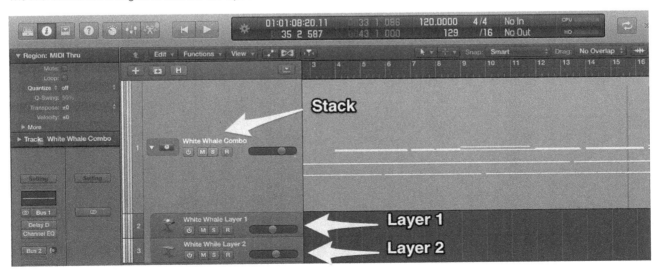

Figure 8.35
The compressor inserted on the Sculpture layer.

Effects and Final Touches

Patches that require constant variation and that have rhythmic features can benefit greatly from the addition of effects such as reverb, delay, stereo imagers, and so forth. You can really improve a combined patch by using some audio effects. Because we used a track stack to combine the two synthesizers, we have full control over which effects we are going to apply to each layer. On the first layer we applied a classic VCA compressor (which is a replica of the DBX 160 compressor/limiter) to reduce some of the peaks in the dynamic generated by some of the variations in the resonance of the filter in Sculpture. To add a bit more "byte" to the layer we also used a soft distortion in the compressor (Figure 8.35).

Figure 8.36
Insert effects on the second layer.

Physical Modeling

Figure 8.37
Delay (top) and Reverb (bottom) on the combined stack.

On the second layer, the one with the more rhythmic part generated by String Studio, we added two effects: an exciter and stereo spreader (Figure 8.36). The former helps us add a bit more presence and character to the patch, and the latter opens up the stereo image of the rhythmic pattern, creating a wider stereo experience.

Finally, on the combined stack of the two layers we added a delay as insert and a 3.0-s reverb as Aux send (Figure 8.37). These two effects help position the patch in a nice large environment that contributes to giving the impression of a bigger and more grandiose sound.

Listen to Audio Example 8.20 to hear the final combined patch with the addition of all the effects.

As we mentioned several times earlier, feel free to experiment with the effects. For these types of creative patches, the sky is really the limit. Let your imagination run free!

Exercises

8.1 Using Table 8.4 as a reference, create the following patches using your physical modeling synthesizer of choice:
- percussive pad
- pads
- ambient/cinematic

8.2 For each of the patches created in the previous exercise, write, sequence, and produce a short 30" cue that best utilizes the sonic characteristics of each patch.

9

Wavetable Synthesis and Granular Synthesis

Introduction

Back in Chapter 7 we explored sample-based synthesis, a technique that for the first time brought with it the allure of faithfully reproducing the sound of acoustic instruments. With the increase in drive storage space, RAM capacities, and processor speeds—not to mention improved techniques for capturing and organizing samples—sample-based instruments increasingly approached an impressive realism. The trade-off, however, was limited facility for crafting the timbre into variations expected in a musical performance. For anyone approaching synthesis with the goal of reproducing the sounds and subtleties of an acoustic instrument, it was still hard to beat what sampling had to offer. What was needed was having that realism coupled with the deep timbral control that was at the heart of synthesis. Additive synthesis would evolve beyond its humble beginnings to address that need in a significant way. The next step was to start with a sample and not just play with its overall harmonic content but to divide it into many slices over its duration and manipulate how and when the slices were reproduced. This is the basis of two similar but different forms of syntheses, wavetable synthesis and granular synthesis.

Both wavetable and granular work on the principle of dividing up a sample into slices over time. With wavetable the slices (commonly referred to as *subtables*) are relatively coarse—older instruments would have 32 divisions over a sample's duration, newer software-based instruments may go as high as 256. The slices of a granular sample are called *grains* and can be as small as 2–50 ms each. As already stated, the two techniques share similarities but there are distinct differences that we will now explore.

Wavetable Synthesis

The beginnings of wavetable synthesis can be traced to the 1980s when Wolfgang Palm introduced the PPG Wave series of synthesizers. His instruments were notable for the rich pads they could produce from wavetables of complex timbres fed into an analog envelope and filter (the sound of digital filters at the time were far less inspiring than their analog counterparts). When the company ceased operation later in the same decade, Waldorf Music adopted the technology and incorporated it into their own designs. Today,

Figure 9.1

Waldorf's Wave 3.V Plug-in (left) and Nave for iPad (right).

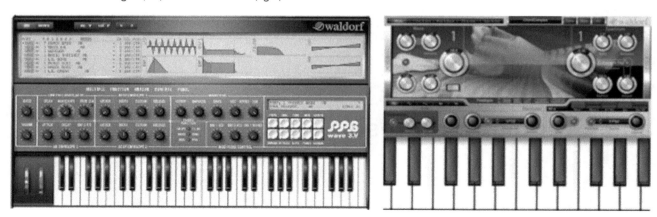

emulations of the original Wave series exist as software as plug-ins and Waldorf's Nave instrument for iPad (Figure 9.1).

Wavetable Types

The term "wavetable" has been used as a label for different synthesis techniques (some more appropriately so than others), leading to confusion over what it is, exactly. In this section we cover the most common types and explain how they work.

Lookup Table

In the 1980s, manufacturers saw a potentially huge market for inexpensive digital keyboards for the home. The need for decent sounds on hardware with limited processing and storage capacities led to incorporating single-cycle complex-wave samples with limited or no modifiers. Selecting patches simply meant pulling a wave cycle from a *lookup table* of options. This has been referred to as wavetable synthesis but should more appropriately be called lookup-table synthesis. This approach and terminology would later be used in inexpensive PC sound cards for the same advantages of cost versus function. Lookup-table synthesis even found its way into the first polyphonic ringtones adopted by Nokia (developed, interestingly enough, by Thomas Dolby's company Beatnik).

Although sometimes tarnished by its lowbrow reputation, a more refined and useful version of lookup tables is still in use within high-quality instruments like Logic's ES2 (see Figure 9.2) on which Apple implemented "Digiwaves" alongside the more familiar sine, triangle, pulse, and saw, and on NI's Massive, which is an advanced synth with many wavetables in each of its oscillators acting as its raw material.

NI's Massive (seen in Figure 9.3) is a highly evolved adaptation of the lookup-table model. The overall structure suggests a modern take on subtractive synthesis with a plethora of modifier capabilities, but clicking on the wave selector in each of the three oscillators reveals an extensive list of wavetables organized under relevant category headings. We have seen this before. What are novel, however, are the controls seen in the oscillator section (image in Figure 9.3, details in Table 9.1). Within each wavetable are 2–200 separate waveforms that are strung together to form a contiguous morphing of the sound. "Wavetable-Position" is a control for selecting a starting point within that wavetable.

Figure 9.2

Logic users may not be aware that the ES2—on the surface a subtractive instrument with an extensive array of modification capabilities—includes a somewhat hidden list of "Digiwaves" in each of its three oscillators accessible by right- or control-clicking on "Sine" (6 o'clock position of the selector dial).

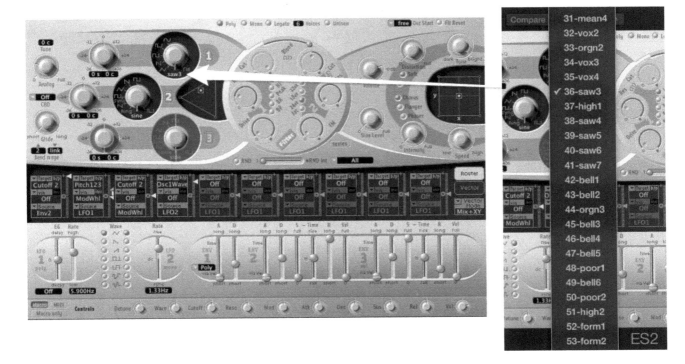

These controls get to a deeper level and reveal the parameters that are now more associated with wavetable synthesis—specifically, having a single cycle *or* a longer sample up to 5 s (optimally) that is synthesized and divided into many time-based segments referred to as "grains." This process shares similarities with granular synthesis but, as we will see, there are some differences.

A common configuration seen in the screenshots of both ES2 and Massive is an organization of similar sounding waves grouped closely together in the lookup tables. This enables coherent shifting up and down the list dynamically using automated controls like LFO and EG, or manual controls like a keyboard's modulation wheel. In Massive, those same controls can be applied to shifts within the 2–200 waves that are contained within each wavetable.

Vector Synthesis

The term vector synthesis (VS) came from Sequential Circuits with the introduction of their Prophet VS (a follow-up to the Prophet-5, the first truly polyphonic synth). The instrument shipped with ninety-six preset waves and thirty-two slots for user-definable waves. The term "vector" refers to a new interface that gives the user control over the selection and blend of four waves—A, B, C, and D—using a joystick (Figure 9.4, Audio Example 9.1).

In addition to the manual control over blend, the joystick position could also be realized (if not actually moved physically) through routing of LFOs and EGs, velocity and pressure from the keyboard, and modulation wheel position through the modulation

Figure 9.3
Native Instruments Massive Synth with blowup of wavetable list and the oscillator section.

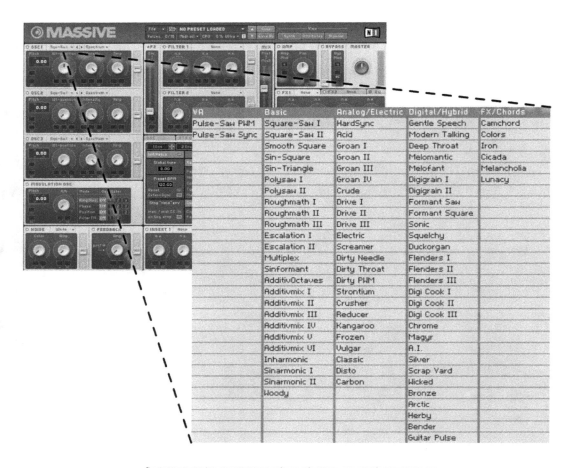

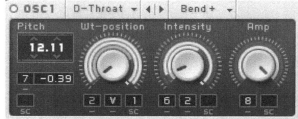

matrix (Figure 9.5, left). A real-time shift in blend could also be defined as part of the path by selecting four different locations within the two-dimensional (2D) plane of the joystick (see Figure 9.5, right) and setting the time duration for moving from one to the next. At its core the VS was still lookup-style wavetables but the timbral results were fresh and dynamically interesting because of these controls.

In the late 1980s, Sequential Circuits went out of business and was picked up by Korg, who would adopt and expand on this technology with their Wavestation synthesizer (see Figure 9.6). The joystick and vector concept would live on with expanded functionality that included *wave sequencing*, which allowed arranging waves linearly for constructing rich evolving pads and interesting rhythmic patterns.

Linear Arithmetic Synthesis

Roland's answer to Korg's success with the Wavestation was the D-50 (Figure 9.7) using what they dubbed linear arithmetic (LA) synthesis. Whereas the Wavestation sequenced

Table 9.1 NI Massive Oscillator Controls Detail

WT position	This wavetable position control reveals a hidden feature of Massive's wavetable: Each selection from the list has from 2 to more than 100 waves that change gradually in character. Turning this knob counterclockwise to clockwise seamlessly transitions through range of options. With manual or dynamic control (LFO, envelope, etc.) over time you can think of this as a two-dimensional vector synthesizer within each oscillator!
Intensity	Tied to the pop-up list of oscillator modes to the top-right of the knob, intensity controls how much the wave is mangled or distorted from the starting point in the wavetable. You really have to hear this to understand what's happening.
Amp	Output volume of the oscillator

Figure 9.4
Detail of vector joystick control of Arturia's Prophet-V (emulation of Sequential Circuits' Prophet 5 synth).

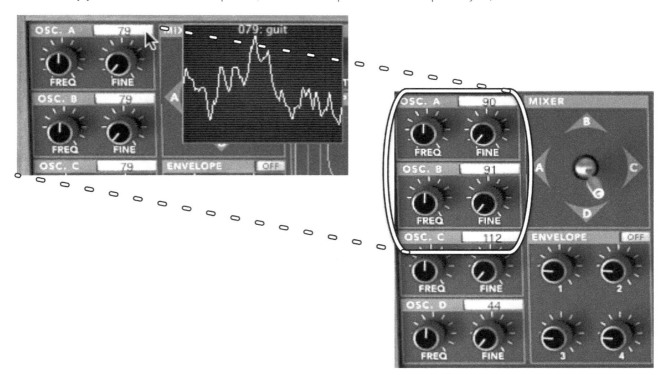

Figure 9.5
Modulation and joystick automation windows in Arturia's Prophet-V.

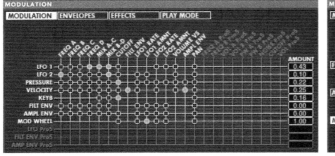
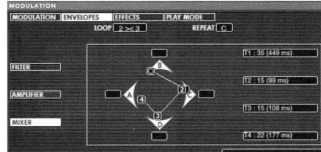

Creating Sounds from Scratch

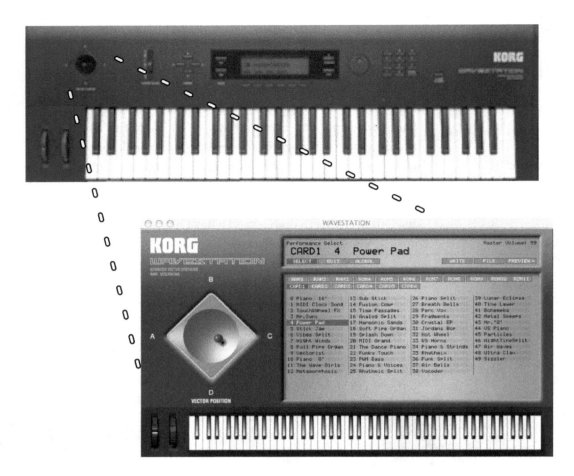

Figure 9.6
Korg Wavestation (top), their first product to incorporate vector synthesis following the acquisition of Sequential Circuits. Korg Wavestation software synthesizer screenshot (bottom) shows the joystick control and patches from the hardware. (Wavestation photo courtesy of Korg USA Inc.).

only wavetables over time, LA instruments used a *sample* for the attack portion of the patch followed by a series of wavetables.

Psychoacoustically it is the attack stage that is critical for our identification of a sound, so the D-50 was particularly good at synthesizing strings, winds, and choir sounds. Using samples for the attack was key to achieving an improved level of realism, and the need to last for just a short period of time meant minimal demand for storage capacity at a time

Figure 9.7
Roland D50 synthesizer: Short samples for the attack stage spliced into .lookup tables that can be modified with subtractive synthesis-style controls (photo used by permission from Roland Corporation US).

when RAM was still costly. Incidentally, it was also the first instrument to incorporate onboard digital effects, giving more depth to the sounds without the need for outboard studio effects.

"Advanced" Wavetable Synthesis

The concept of wavetable synthesis evolved from the previous implementations we just covered into something more complex and useful for sound designers working on a DAW rather than on a hardware instrument. In general, plug-in instruments continue a trend of blurring the lines between the different synthesis techniques covered in this book, and wavetable is among them. "Advanced Wavetable Synthesis" is not a common term in synthesis parlance but one that we are using to describe new advances in wavetable that borrow much from what we will see had been the sole domain of granular.

Wavetable instruments still have wavetables to use as a starting point but a new function known as *resynthesis* allows importing your own samples that are divided into time slices called *grains*. The advantage here is that grains can be sequenced and played forward or backward, the speed of playback through the grains is variable, and start, end, and loop points can be defined within the overall sequence—all characteristics of granular synthesis but with fewer and larger grains.

Figure 9.8 shows a sound sample in a standard amplitude/time graph waveform, a spectrogram showing the sample's amplitude, frequency, and time, and a spectrogram revealing the individual grains after the sample has been resynthesized. Listen to Audio Example 9.1 to hear the stages of this progression.

The resynthesized version, as expected, does not sound exactly the same as the sample but the source is clearly evident. It is a re-creation of the original sample with sixty-four grains spanning the sample's duration representing changes in the frequency spectrum at regular time intervals, collectively called the "wavetable." Relating back to Massive, as an example, when a wavetable is selected in one of the oscillators, it is already recalled in a form that is similar to what we see in the Figure 9.8's "resynthesized version." Additionally, when Massive's "Wt-position" knob is rotated, a different grain in the sequence is selected as the starting point. The image in this example is taken from a custom sample loaded into Waves' Codex, an instrument that is called a wavetable synthesizer, but it has elements of subtractive, FM, and granular. Let's look closer at Codex as a case study of the parameters common to this more advanced category of wavetable synths (see Figure 9.9).

Figure 9.8
Example of sound sample resynthesized into grains.

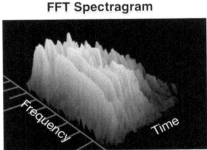
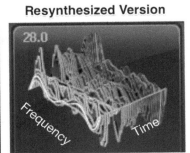

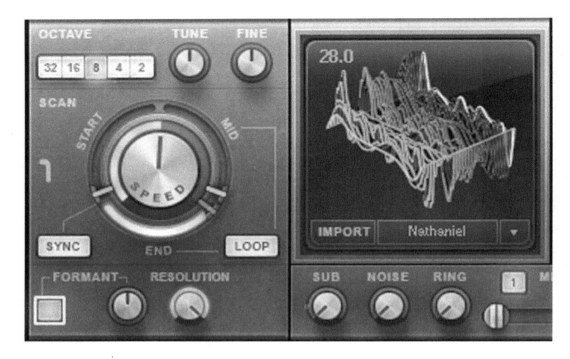

Figure 9.9
OSC 1 in Waves' Codex Wavetable synthesizer.

The large dial on the left-hand side of the spectrogram offers control over the START (starting grain, which now is set to the beginning but can be shifted completely through to the end of the wavetable), the MID (midpoint, where a loop can be defined between that location and the end point), and the END (end point). Note that if the START location is *after* the END location the wavetable will play backwards! If the START and END points are close together and looping is enabled, a single-cycle wavetable (akin to older instruments) is possible. SPEED, as you may have guessed, affects how quickly the wavetable is played from the START to the END locations. And because SPEED controls only how long each grain is played, there is no effect on the pitch.

We are quickly getting into territory that has traditionally been considered granular synthesis, but before going there let's look at Table 9.2 in which we have made an attempt at defining distinguishing characteristics of each.

Granular Synthesis

Granular synthesis is really where the gap between synthesis and sampling is finally closed. Synthesis in all its prior forms offers unique, unnatural timbres and creative manipulation tools but it could not approach the realism that became possible with sampling. Sampling of some sounds (percussion in particular) has gotten so good that it can be hard to tell the difference between a sample and a recording of a real instrument, especially when embedded in a track and "sweetened" with signal processing. Sound modification, however, was limited to tools that were familiar to anyone with experience in subtractive synthesis. By starting with a sample but—instead of simply playing it back—dividing it into very small grains, a sample at normal speed can be played back largely intact. Shifting playback speed, direction, and order of grain playback retains an organic association with the original sound while producing something

Table 9.2 Where wavetable synthesis ends and granular synthesis begins is challenging to define, but here is an attempt to create a distinction.

	Wavetable	**Granular**
Raw material	Single-cycle preloaded waves	Prerecorded samples
	Resynthesized user-recorded samples	User-recorded samples
		Actual samples, not resynthesized
Grain composition	1–100 grains per wavetable	1500+ grains for 5-s sample
	Duration: 10–100 ms	Duration: 3–5 ms
Grain sequence	Grains can be played in order, either forward or backward	Grains can be played forward or backward in order
	Midpoint can be added to define loop area	Grains can be played in random order in a "grain cloud"
Layering	Vector synthesis commonly allowed blending of up to four wavetables	Typically one sample at a time, but many grains (as many as 1000 in some software) may be overlapped in a "grain cloud"
Processing Demands	Moderate	Heavy

new. Granular synthesis works with a sound file (not a resynthesized sample, as in wavetable synthesis) divided into lots of really small grains, typically no more than 3–5-ms duration.

In fact, stepping away from synthesis for a moment, granular technology is at the core of many time-shifting algorithms that are part of all modern DAWs. Real-time tempo shifting is possible because vocal and recorded instrument tracks are sliced up into fine grains. Shifting the start point of each grain closer together and crossfading to avoid artifacts is an extremely efficient method for speeding up a tempo and not changing pitch. Slowing down from the recorded tempo adds the challenge of spreading the grains and thus creating gaps that must be filled with synthesized segments created as an extrapolation from what came before and what comes after.

Reason is a popular platform for composing and sound design and includes several powerful instruments, one of which, Malström, is referred to as a "Graintable" synthesizer (as if we needed more evidence of the merger of these techniques) (see Figure 9.10 and Table 9.3).

Because of intense processing demands, granular synthesis was for most of its lifetime the sole domain of composers and sound designers with access to powerful computer systems housed at a college or university. However, personal computers would ultimately achieve sufficient capacities required for the complex, real-time processing of granular synthesis, and today there are even sophisticated instruments available on tablets and smartphones. These and other powerful wavetable and granular instruments as well as recommended techniques are explored in The Producer Point of View.

Figure 9.10
Reason's Malström with focus on oscillator.

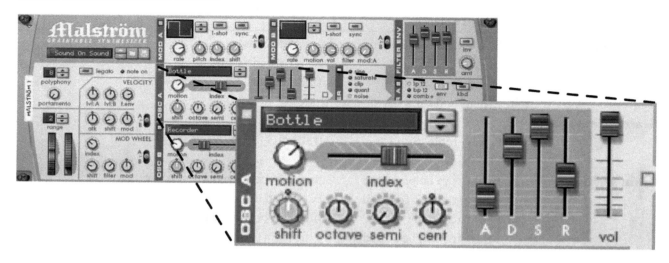

Table 9.3 Details on Granular Synthesis-Specific Parameters in Malström

Motion	The speed the grains are played back, independent of pitch
Index	The slider represents the length of the grain table; moving the slider selects the start location of playback
Shift	Adjusts the sound's format without affecting its pitch

The Producer Point of View on Wavetable Synthesis and Granular Synthesis

From a producer point of view, wavetable synthesis is extremely exciting. It allows us to step into some very creative territories, in which the boundaries are basically confined only by our imagination. We always feel that something unexpected and surprising can be created when programming a new patch! Wavetable is well suited for interesting and evolving pads, leads, and sound effects. In recent years it has become very popular, particularly in electronic and electronic-pop music. Among some of our favorites wavetable hardware synthesizers are the Waldorf Wave (and its younger "cousins," the Microwave family), and the Korg Wavestation. These machines were really the first popular wavetable synthesizers used for commercial music productions. Artists like Enya, Depeche Mode, and Deep Forest have used extensively these machines in several of their productions. The sonic characteristics of these early wavetable synthesizers were pretty exceptional, ranging from dreamy pads to punchy basses, from electric piano-like patches to string-based instruments, from aggressive and sweeping pads all the way to sci-fi effects. The versatility of the wavetable engine is remarkable. As we learned in the first part of this chapter, the ability of creating sonorities that change over time following both predictable and unpredictable patterns allows us to set our creativity free. What makes this synthesis engine particularly interesting is also the ability to feed our own waveforms and tables, therefore giving us the options of almost infinite combinations. For example we can import a vocal sample, blend it with a second table, and then have a complex matrix of LFOs creating interesting variations over time. Another technique that we find very interesting to use and that we are going to learn in this second half of this chapter, is to sample an analog synth (like a Roland Juno 106), import the sample in our wavetable synthesizer, and then blend it with a digitally generated wave. This allows us to create a hybrid patch that maintains the "fattiness" of the analog source, but, at the same time, features the versatility of digital controls and modulators. Let's take a look at some of the sonorities that can be more effectively created and applied to contemporary production using wavetable synthesis (Table 9.4).

Wavetable Software Synthesizers for the Contemporary Producer

There is a variety of tools available when it comes to wavetable synthesis. A technique that just a decade or two ago required big and heavy hardware synthesizers now can be easily incorporated into your production with a laptop or even a tablet. There are two main categories of wavetable software synthesizers available. The first one features a more modular approach, which can be more powerful and versatile, but also more complex and sometimes more time consuming to program. Synthesizers such as Reaktor or Tassman fall into this category. The second option features wavetable synthesizers that are more user friendly and faster to program, without sacrificing versatility and creative power. Among them we really like Codex (Waves), Serum (by Xfer Records) (Figure 9.11), and Nave for iPad (by Waldorf). These synthesizers, extremely powerful and versatile, but without the complication of convoluted interfaces, are the ideal tools for the contemporary electronic composer and producer.

Wavetable sonorities can be used in a variety of contemporary styles and productions including Dubstep, Trance, Acid, and Chemical, for which original and aggressive sonorities are needed, as well as in more mellow genres such as Ambient Techno and Ambient Trance, where smoother and changing pads are featured. Let's take a look at how to create a couple of signature patches for contemporary productions using a wavetable synthesizer!

Creating an Aggressive Pad Patch With a Wavetable Synthesizer

Our first patch is going to be an aggressive pad. In this example we are using Waves' Codex. Keep in mind though that the same procedure and structure we are going to learn in Codex can be easily applied to other wavetable software synthesizers. The first thing that we need to do is to initialize the setting of our wavetable, by selecting the "Init" patch (Figure 9.12). This will allow us to start from a blank slate.

One of the most important aspects of creating a new patch in a wavetable synthesizer is the choice of the waves that form the main table. In most software synthesizers we have two oscillators available, each with the capability of holding a series of waveforms combined in a table. We can usually choose from a preset list of tables containing between 64 and 128 waveforms (Figure 9.13).

As you can see, the selection of preset tables is fairly comprehensive, and you can most likely find anything you need to get started with the creation of some really interesting patches. One of the aspects that we really like of creating original sonorities with

Table 9.4 Wavetable Synthesis Patch Characteristics

Category	Sonic Features	Compositional and Production Features
Dreamy/light pads	Light and airy Liquid texture Slow attack and release Evolving slowly	These pads tend to be light and soft, with cyclic and smooth overtime variations. The automatic swipe between different waves of a table creates the effect of a constant variation.
Aggressive pads	Edgy and rough Evolving time medium to fast Short to medium attack Medium to long release	Aggressive pads can be generated starting from more traditional geometrically shaped evolving waveforms. The shorter attack provides the necessary aggressiveness, whereas the more "analog" basic waveform adds to the grunginess of the patch.
Sweepy pads/leads	Fast sweeping filter Bright and edgy Complex LFO and modulation routing	The fact that wavetable synthesis allows for a continuous shift between contiguous waveforms makes it an ideal tool for creating pads and leads that feature continuous sweeps.
Electronic bass	Short attack Fat or edgy character Used with arpeggiator/sequence generator	Wavetable synthesis seems to have a special role in creating powerful bass patches. Using more traditional geometrically shaped waveforms or importing some sampled waveforms from legendary analog synthesizers, and adding a complex LFO matrix allow us to create some interesting bass sonorities.
Rhythmic pads	Cyclic table switching Use of arpeggiator Use of LFOs synced to the host tempo	Switching between waveforms of a table using either a synced LFO or a step sequencer/arpeggiator can create excellent rhythmic/sequenced pads or basses
EFX	Based on original wavetables Complex LFO and modulation routing Heavy use of audio effects such as modulation, delay, distortion, and reverb	This is one of our favorite features of wavetable synthesis. Because we are able to import complex waveforms and therefore create our table, the sky is really the limit. Importing voice-based samples can be a great starting point to generate some very original sonorities.

Figure 9.11
Wavetable software synthesizer Serum by Xfer Records.

Figure 9.12
Selecting the "Init" patch in Codex.

Creating Sounds from Scratch

Figure 9.13
Some of the preset tables available in Codex.

wavetable synthesis is the ability to import your own tables. In general you can import any type of sample. Try to keep the length of the imported sample between 1 and 5 seconds to get the best results. In particular we like to import samples of analog synthesizers, recorded from the original hardware machines. This technique allows us to work on some really rich sonic material that can be manipulated to extreme levels through wavetable synthesis. For this aggressive pad patch we are going to use two original samples from two legendary vintage hardware synthesizers: a Roland Juno 106 and a Yamaha CS-10 (Figure 9.14).

To import the samples click on the "Import" button in each of the two table widows (Figure 9.15) and navigate to your sample folder. Listen to the original two waveforms recorded from the analog synthesizers (Audio Examples 9.2 and 9.3) and compare them with the resynthesized versions after we imported them in Codex (Audio Examples 9.4 and 9.5).

When selecting the raw material for your patch try to choose tables that are somehow complementary to each other. For example, you might want to have a table that features a

Wavetable and Granular Syntheses

Figure 9.14
The "real deal": Roland Juno 106 and Yamaha CS-10.

strong attack, while the other one has a long and evolving sustain; or a low and deep table could be paired with an edgier and thinner one.

Beginning and End

Once we have selected the table, we need to select the start, end, and range of each table. Wavetable synthesis allows us to swipe and morph into each single wave of each table. For the pad that we are designing, we want to make sure that we keep swiping for the entire length of the table in order to have a continuous changing effect. Pay particular attention

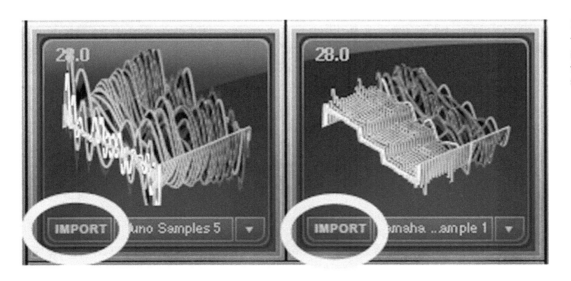

Figure 9.15
The two tables in Codex with the imported samples.

Creating Sounds from Scratch

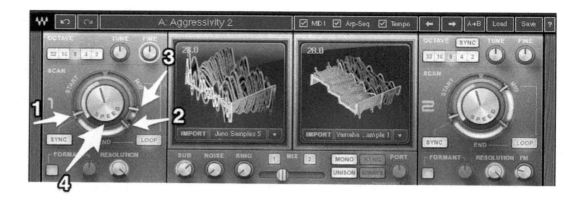

Figure 9.16
The two tables in Codex with the start (1), end (2), mid (3), and speed (4) parameters for the first table highlighted.

to the start of the table because this will also have an impact on the attack of the sound. For an aggressive pad we want to keep the sharp attack of the original analog sample of the Juno 106, and therefore we can have the start set right at waveform 1 of the table. Because we want to keep the full shape of the original sample we have imported, we also set the end point to the last waveform of the table (Figure 9.16). This will guarantee that we will be able to hear the full length of the original sample.

To add more motion to the pad, we recommend looping the table so that when the end is reached the table is loped back between the midpoint and the end point. This will allow us to have a constantly moving patch. When setting the midpoint and the end point, make sure to select two waveforms that create a nice and smooth loop. For example, if the table starts with a sharp attack, we like to move the midpoint a bit farther in the table so that when looping the table the attack is not included in the loop area. Repeat the same process for the second table. As you can see in Figure 9.16, for the second table (the one on the right) that has the Yamaha sample assigned, we moved the midpoint to the middle of the table to provide a more regular loop effect. The way a wavetable synthesizer goes through a table is determined by the "Speed" parameter. In general we prefer trying to match the speed of the original sample, but here you can experiment and be creative. A slower speed will work better for smoother sonorities that need to give the impression of a slow-over-time variation, whereas a faster speed will give the impression of a more cyclic change, adding more energy to the patch. For more interesting pad sonorities, try to have the two tables moving at different speeds to create more variation. You do not always want to use the entire table for a patch. In effect, if you want a sonority that is a bit more stationary, you can set the speed parameter to zero so that the table will play only the wave assigned to the start position. This technique is particularly effective if used on one table only. Pick a specific wave for the first table while having the second table goes through all its waves. The first static table will mainly control the color of the attack of the patch, and the second table will determine the sustain portion of the patch.

Listen to Audio Examples 9.6 and 9.7 to compare a pad programmed with both tables moving among the different waves and a pad that features the first table as stationary.

Balancing

Once you have the two tables set, spend some time setting the balance between the two oscillators. If the two tables are sonically complementary then you can really use the mix

Figure 9.17
The mix balance between the two oscillators in Codex.

parameter to nicely balance between the attack and the sustain parts of the patch. In the case of the aggressive pad we used a bit more of the Juno samples, which have sharper attacks (Figure 9.17).

To add a bit more character to the patch, we also recommend mixing a bit of the sub-oscillator and the ring modulator (Figure 9.17). The global harshness of this pad patch can be controlled with the "Resolution" parameter in Codex. Each oscillator can be used at full digital resolution (100) or at a lower one. Lowering this parameter allows us to add crunchiness and a characteristic "byte" to the patch.

A Bit of Filtering

Once you have worked out the settings for the oscillators, it is time to adjust the VCA and the VCF sections. As we learned earlier, for a pad generally we are looking for a mid-to-long release and a short-to-mid attack. Because we are looking for a more aggressive pad we recommend leaving the attack of the amplifier set to its minimum position (Figure 9.18).

To accentuate the attack portion of the patch and highlight its edginess, we used a HPF with a fairly low cutoff frequency. The added resonance creates a nice extra punch in the mid–low frequencies. Its slightly longer attack on the filter envelope creates a more interesting sonority during the first half of the patch. Listen to Audio Examples 9.8 and 9.9 to compare the patch without and with the filter/resonance applied, respectively.

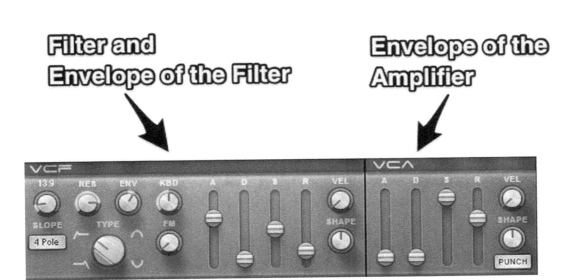

Figure 9.18
The filter, the envelope of the filter, and the envelope of the amplifier in Codex.

Creating Sounds from Scratch

Figure 9.19
The modulation matrix in Codex: the oscillator tables assigned to LFOs.

Modulation Matrix

What makes a modern wavetable software synthesizer particularity appealing when creating sophisticated patches is the advanced option of a complex modulation matrix. In this section usually we can control any parameter and aspect of our patch via LFOs or external controllers.

Although the combination of an advanced modulation matrix can be potentially infinite, there are a few options that we like to apply when creating pad patches. To take full advantage of the main characteristic of a wavetable synthesizer, we like to assign both oscillators' tables to a separate LFO. For the pad we are working on, we assigned the table of OSC 1 to LFO 3 and the table of OSC 2 to LFO 2. As you can see in Figure 9.19, LFO 3 features a random waveform with a synced time parameter of sixteenth notes. LFO 2 is set to a square wave with a free rate (not synced to the sequence tempo).

Having the oscillator table controlled by an LFO means that the table will be switching, at the selected rate, between the current waveform in the imported table and the selected waveform of the LFO. This creates an interesting variation inside the chosen table, giving the effect of a waveform that is constantly changing over time. We recommend usually assigning one LFO to a beat-based rate and one to a freely assignable rate to create more variation between the two oscillators. Of course it is possible to take advantage even further of the modulation capabilities of our wavetable synthesizer to add more character and originality to our aggressive pad patch. In particular you can use another LFO to control the scan start point of one of the two oscillator tables (Figure 9.20).

Figure 9.20
The modulation matrix in Codex with the LFO 4 controlling the Scan start of oscillator 1.

Try to set the rate of the LFO (number 4 in this case) with a square wave and a rhythmic subdivision of 1/8 or 1/16. The resulting sonority will be more rhythmic and pulsating.

Listen to Audio Examples 9.10 and 9.11 to compare the pads we created without and with modulations, respectively.

Assigning the modulation wheel to the LFO 4's rate will allow you to control the speed at which the scan start point changes, adding more control over the rhythmic activity of the pad (Audio Example 9.12).

Effects

As we saw for the other types of synthesis, the addition of audio effects and/or an arpeggiator can really bring your patch to another level of sophistication. For any aggressive patches (such as pad, bass, or lead), you should try adding low to medium distortion or saturation. Some wavetable synthesizers offer built-in effects that can be convenient for you to use when programming original sounds. In the case of the aggressive pad we are building for this example, we used a subtle distortion effect available directly in Codex. For pads and leads in general, adding a bit of reverb and delay can help thicken the patch. Be careful, though, not to overuse them because a "wetter" sonority can sound mellower and less edgy. Finally, you can add a bit more modulation effect such as chorus or flanger to open up the stereo image and to add a bit of depth (Figure 9.21).

Listen to Audio Examples 9.13 and 9.14 to compare the aggressive pad we programmed so far without and with the aforementioned effects applied, respectively.

Rhythmic Pads With the Arpeggiator

If you plan to assign to the pad a more rhythmic function inside your piece, we recommend using an arpeggiator. Some wavetable synthesizers feature a built-in arpeggiator that can deeply integrate with some of the modulation matrices. We like to use the arpeggiator for mainly two purposes: a pitch-based function and a modulation function.

Figure 9.21
The effect section in Codex for the aggressive pad patch.

Figure 9.22
The arpeggiator in Codex with a melodic function.

Figure 9.23
Music notation of the pattern shown in Figure 9.22.

The former allows us to control the pitch of the patch and therefore create a pattern based on the step sequencer philosophy. Each step of the arpeggiator can be assigned to a different pitch, creating a rhythmic/melodic pattern that can be triggered and transposed through the note currently being played. In Figure 9.22 you can see the arpeggiator in Codex triggering a sixteen-step pattern in which some of the steps are transposed by semitones.

As you can see, the arpeggiator has a sequence of sixteen notes. Each step can be transposed up (+) or down (−) with a resolution of 1/2 steps. The patterns represented in Figure 9.22 can be transcribed using music notation, as shown in Figure 9.23.

Listen to Audio Example 9.15 to hear this arpeggiator pattern programmed in Codex.

To highlight the rhythmic aspect of the arpeggiator we can also reduce its "Gate" parameter (therefore making each step shorter and more staccato) and lower the resolution of the two oscillators (Audio Example 9.16).

As we mentioned earlier, the arpeggiator can also be used as part of the modulation matrix and therefore used to control a parameter of the patch. Try to assign it to the resonance of the VCF to create more variations based the rhythmic pattern played by the arpeggiator (Audio Example 9.17).

As you can see, the possibilities are pretty much endless.

Electronic Bass Using Wavetable Synthesis

Another category of sounds that can be interesting to create for any electronic production using wavetable synthesis is that of electronic basses, specifically, basses that feature a medium-to-high level of punchiness and grunginess. When creating these types of patches we recommend thinking outside the box and starting with some original waveforms. For example, we like to use vocal samples (such as a short phrase) as the basic material for one of our tables. For this bass patch we recorded a simple phrase, "More cowbell please!" and then imported it into oscillator 1's table. Listen to Audio Examples 9.18 and 9.19 to compare the original sample and the one resynthesized by Codex.

The vocal samples are very flexible and can be used to bring the patch in different directions. To add a more pitch-based component to the bass patch, for the second

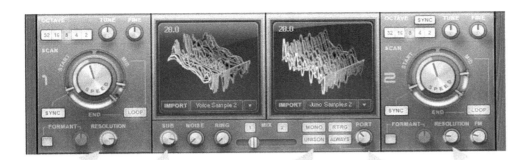

Figure 9.24
Oscillator tables settings for a Wavetable bass patch.

oscillator's table we imported a sample from the Juno 106 that features a short attack and a slightly sweeping sustain (Audio Example 9.20).

Set the start points, midpoints, and end points of the two oscillator tables to your liking. Make sure to have enough "punch" and sustain. For this example we picked a fairly standard setting (Figure 9.24).

Lower the resolution for the two oscillators to get a "punchier" and "dirtier" sound (Figure 9.24). To have a slightly fatter and present sound, turn the "Unison" option on. This button adds a doubling effect, which creates a richer sonority. The "Mono" option sets the synthesizer as monophonic, and it is particularly indicated for a bass patch because it allows us to play only one note at the time and, even more important, to use the *portamento* (Port) knob. We find that, particularly for electronic bass patches, *portamento* is an essential feature that adds more expressivity and makes the patch more interesting. For a bass patch we suggest also adding a bit of the "sub" oscillator in order to obtain a deeper effect.

Listen to Audio Examples 9.21 and 9.22 to compare a bass patch without the suboscillator and without the unison option engaged, respectively, and the same patch with both options enabled.

For a bass patch of this type, set the VCA with a short attack and release in order to maximize its rhythmic possibilities. For the VCF we like to use a LPF in order to maintain the low frequencies of the patch with a fairly high cutoff frequency and a good amount of resonance (Figure 9.25).

Combining Arpeggiator and Modulation

For a bass patch that has the requirements of being aggressive, edgy, and always captivating, the modulation matrix is the place for you to let your creativity run free! As we described earlier in this chapter when discussing the aggressive pad patch, we like to use LFOs to control parameters

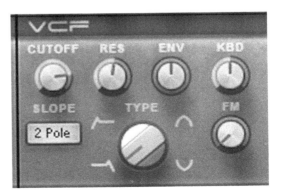

Figure 9.25
VCF settings for a bass patch.

Creating Sounds from Scratch

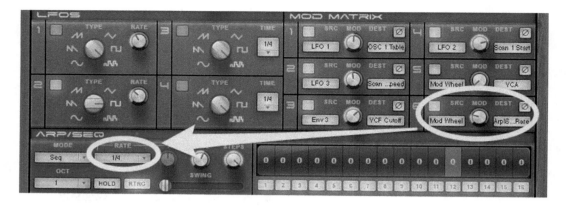

Figure 9.26
Controlling the arpeggiator rate via the modulation wheel.

such as tables, scan speed, start, and mid positions, or the resonance of the filter. All these options are excellent for creating sonorities that change over time and are always fresh and interesting. One type of modulation, though, that we find particularly interesting for an electronic bass patch is the one that allows us to control a parameter of the arpeggiator via a MIDI CC message. Specifically we can control the rate of the arpeggiator via the modulation wheel of our keyboard controller. This technique allows us to create interesting rhythmic patterns that are typically found in Dub Step. The idea is that we set one of the modulator slots in the modulation matrix of our wavetable synthesizer to be controlled by a MIDI CC message (in this case number 1 Modulation). The target will be the rate parameter of the arpeggiator (Figure 9.26).

Listen to Audio Example 9.23 to hear the effect of the modulation wheel controlling the rate of the arpeggiator in Codex.

Audio Effects

Electronic bass patches can also greatly benefit from the addition of audio effects such as chorus, saturation, light to medium distortion, amp simulators, and compressor. Using these effects, you can create a large number of variations of the same patch. If you are looking for an added low-end effect, try to add an extra sub-bass effect with the addition of a vintage compressor (Figure 9.27).

Listen to Audio Example 9.24 to hear the bass patch with the MaxxBass and compressor applied.

To further thickening the bass sound you can use a chorus effect. Although this will smooth out some of the edginess of the patch, it will also improve its spatial placement by opening its stereo image (Figure 9.28).

Listen to Audio Example 9.25 to hear the bass patch with the MondoMod (Waves) chorus applied.

If you need to add more byte and character to the bass patch, try using a distortion effect (Figure 9.29) with a medium drive, so that you get the byte of the distortion without completely overpowering the original sonority.

To create a bass patch with more character, we also recommend doubling the bass part an octave higher and delaying the double by thirty to forty ticks. Pan the two parts slightly off the center position in opposite directions in order to widen the stereo image. This is easily achieved with the Stack feature in Logic Pro X (Figure 9.30).

Figure 9.27
Add some extra low-end and punch with MaxxBass (Waves) and a vintage compressor (Logic Pro X).

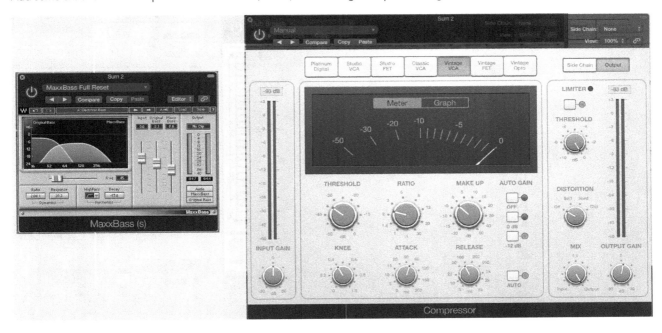

Figure 9.28
Add some chorus effects to open up the stereo image and fatten a bit the overall sound.

Creating Sounds from Scratch

Figure 9.29
Adding some distortion to the bass patch.

Listen to Audio Example 9.26 to hear the bass patch with the distortion and the octave doubling applied.

Granular Software Synthesizers for the Contemporary Producer

Granular synthesis provides some of the more creative options when it comes to sound design and production. Its ability to virtually transform any given audio material (simple or complex) into something totally different is very appealing to any contemporary producer. Although there are some hardware-based granular synthesizers available, most of them are either computer based or just a bit more than prototypes. In general, software-based granular synthesizers are more popular, powerful, and definitely easier to program and interact with. On the software side, the options are many and, in most cases, of excellent quality. Some of our favorites are Padshop by Steinberg, Crusher-X by accSone, Malström in Reason by Propellerhead, and Reaktor by Native Instruments, just to mention a few. They all have their strengths and special features depending on the type of music or sonority you are looking to create. Most of these are relatively affordable or have trial licenses. Therefore we recommend trying them and seeing which interface or feature seems more appealing to your production style.

Real Time vs. Synthesizer

There are two main types of synthesizers based on granular synthesis that you can use for your productions: The first type is based on real-time granular synthesis applied to any

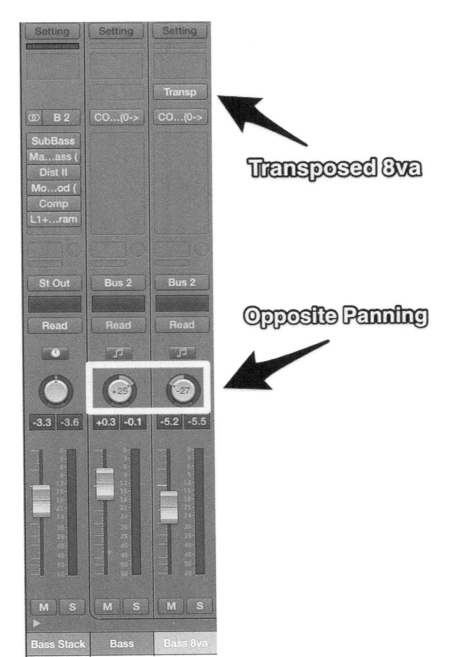

Figure 9.30 Doubling an octave higher the bass part and panning it off center in Logic Pro X.

audio source you decide to process. For example, you can apply granular synthesis to an audio track inside your DAW; this could be a vocal track, a drum/percussive track, and so forth. A typical example of this type of plug in is CrusherX-Live by accSone (http://www.accsone.com/). These are audio applications that can process in real time any audio signal that is fed to their input (Figure 9.31).

These tools are very powerful and can be effective when processing audio during a live performance. They can also be used to transform a non-granular synthesizer (for example, a hardware synthesizer) into a virtual granular synthesizer. To demonstrate the effectiveness of this, we processed a live audio out from a Yamaha CS-10 (Audio Example 9.27) with CrusherX-Live to create a more interesting sonority (Audio Example 9.28).

Creating Sounds from Scratch

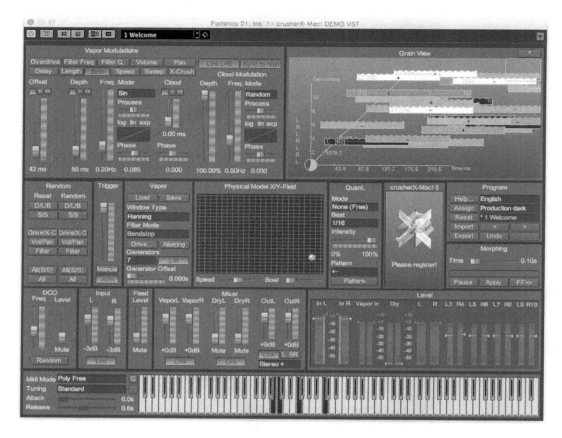

Figure 9.31
CrusherX-Live real-time Granular Synthesizer plug-in.

Real-time plug-ins can also be very interesting when used on rhythmic tracks such as drum or percussion loops. In this particular case, changing the size of the grains can be a creative way to add some interesting twists to a more traditional rhythmic part. Try to add a granular synthesis plug-in to a simple repetitive loop while assigning the size of the grain to a MIDI CC. Experiment with other parameters such as position of the grains, spread, type of loop, and speed of the grains.

In Audio Examples 9.29–9.32 we took a simple drum loop (Audio Example 9.29) and processed it with CrusherX-Live, changing some of the main parameters as shown in Table 9.5.

Granular Software Synthesizers

The second application of granular synthesis is in a more traditional patch-creation synthesizer such as Padshop (Figure 9.32) or Malström, where we work on building granular patches with which we interact via a MIDI controller.

These tools are excellent for creating any type of pads (both pitched and nonpitched), ranging from more traditional soft pads all the way to extreme aggressive or "schizophrenic" beds of sounds, which, in extreme cases, can easily morph into sound effects. Look at Table 9.6 for a list of some of the more characteristic granular synthesis pads that we like to use for different types of electronic productions.

Keep in mind that the one shown in the previous table is a not a complete list of what granular synthesis can provide for a modern producer, but instead a list of our favorite types of patches that can perfectly complement the other types of synthesis we discussed

Wavetable and Granular Syntheses

Table 9.5 Examples of Typical Granular Synthesis Parameters Used in Audio Examples 9.29–9.32

Parameter Changed	Sonic Observation and Alterations	Audio Example
No effect applied	Plain audio file	Audio Example 9.29
Length	The length of the grain is measured in milliseconds. Changing this parameter allows you to morph from "spiky" sounds (very short grains) to more hectic sound collages.	Audio Example 9.30
Speed	The speed at which the grain is played has an effect on its pitch.	Audio Example 9.31
Birth	It controls the distance between grains. The main sonic effect is a choppy waveform with larger empty spaces in between sonic events	Audio Example 9.32

Figure 9.32
Padshop by Steinberg.

Table 9.6 Common Types of Pads Programmed With Granular Synthesis

Type of Pad	Main Sonic Characteristics	Comments
Traditional soft and light	Medium to long attack Soft or "rubbery" with medium dense to light texture Vocal/choir quality Subtle variations over time based on longer grains structures	This category features some of the most commonly used granular synthesis pads, where at the core of the patch there is a vocal sample or a sustained "glass-like" sample. We like these pads for their clarity and richness.
Edgy and sharp	Medium attack Glassy and sharp sonic features Based on shorter grains structures Use of resonance and modulated format to increase over-time variation	This category features sharper, more metallic, and/or glassy patches. The use of shorter grains allows for more abrasive and slightly more hectic sonorities.
Irregular rhythmic	Rhythmic effect based on an imported complex waveform Very creative and original Used for both SFX and rhythmic compositional elements	These sonorities are usually based on loops or rhythmic material imported into the granular synthesizer and then manipulated via time-stretching algorithms.
EFX	Based on imported complex waveforms Short grain structures Random grain and cloud playback Heavy modulation applied to some key parameters such as grain speed, grain duration, format, and filter resonance	This category is the most fun to play with. Basically the only rule is "anything goes!." From ambient ever-changing sonorities all the way to sci-fi cues and horror movie soundtracks. Experiment with small grains, random clouds, speed and direction changes.

in the previous chapters. These patches can easily be used in some of the more contemporary styles in electronic music from Tripop, to Ambient, from Trance to Psybient.

Building Granular Synthesizer Pads

In this section we take a look at how to create an irregular/hectic rhythmic pad and some variations. For these examples we will use one of our favorite production synthesizers: Padshop by Steinberg. Let's start!

In most granular software synthesizers, we have available two generators featuring independent waveforms, LFOs, modulators, envelopes, and effects. When programming a pad that has a specific feature (such as the rhythm aspects for the one that we are working on here), we like to start from two very different "raw" waveforms: one that is intrinsically tight to the main characteristic of the patch, and the other one that is complementary or sometimes even diametrically opposite to the main sonic feature of the patch we are programming. Having two sources that complement each other allows us to create a patch that is more balanced and that can cover a wider sonic landscape. For this rhythmic pad we picked two samples (Layers A and B) that are complementary to each other. The first one is an acoustic guitar played with a drumstick (Audio Example 9.33). This waveform provides a nice rhythmic alternation of fast transients (when the stick hit the strings) and more sustained parts. The guitar covers the mid and high ranges of the spectrum, leaving space in the low-end register for our next generator.

Our second generator (Layer B) features a waveform that doesn't have any rhythmic element per se. It is a smooth sustained arco bass note (Audio Example 9.34). In fact this is one of the samples we used for our acoustic bass patch in the sampling chapter. We picked this waveform for two main reasons: It is the perfect complement to the rhythmic waveform of Layer A, and it is the perfect example of how the same sonic material can be transformed in completely new sonorities using different types of synthesis.

First Generator—Layer A

Once we have the two main waveforms to play with we are going to work with the grain settings for the first layer (Layer A). Set the starting position of the grains inside the imported waveform (Figure 9.33). This parameter determines where the grain(s) starts when triggered. For a percussive patch we like to set it right where a strong transient is (in this case we set it to the beginning of the waveform).

To add additional variations to the position of the grains, try using the "Random" option. Higher values move the start of the grains inside the waveforms. With a setting of 100% the grains span across the entire waveform. One of the most important aspects of how your patch is going to sound when using granular synthesis is to determine the number, speed, and duration of the grains. Basically the process is similar to determining the

Figure 9.33
Grain settings for Layer A in Padshop.

number and type of musicians when you start writing for an ensemble; different instruments will determine if the piece is going to be classical, rock, country, and so forth. For the rhythmic layer (Layer A in Padshop) we are using a single-grain setup (Figure 9.33) with a speed of 100% (original speed of the waveform); this determines the speed at which the grain position progresses through the waveform. If you want the grain to progress in the reverse direction use negative values. Use the Duration parameter to determine how the pitch of the pad will be generated. With very short grains the pitch of the patch will be based on the frequency at which the grains are repeated. With a duration of 10 or higher, the pad will get the pitch of the original waveform. Listen to Audio Examples 9.35–9.37 to compare a grain with short, medium, and long durations, respectively.

For this patch we are using a medium grain duration set to 50 in order to keep enough of the original waveform character without making it too recognizable (Audio Example 9.36). A Random option allows us also to randomize the duration of the grains for added variation.

Another important parameter of a granular synthesizer is the number of grains used in the patch. A low number of grains will result in a sound that is closer to that of the original waveform (Audio Example 9.38), whereas a higher number will move the sound toward a more "chaotic" sonic cloud (Audio Example 9.39).

For this patch we used a single grain for the rhythmic layer, as shown in Figure 9.33.

We also have control over the pitch of the grains forming the sound cloud. The pitch can be offset in semitones (Pitch st) and randomized via the Random parameter (Figure 9.33). If we are using more than one grain, then we can change the pitch of each grain via the Spread parameter (Figure 9.33). For the pad we are creating we are going to leave the pitch unaltered for Layer A, but for more experimental patches you might want to use these settings to create more variations.

Second Generator—Layer B

For the second generator (Layer B) of our pad we picked a very different waveform, an arco bass (pitch G2) sample (Figure 9.34). This is a smooth and sustained waveform. It

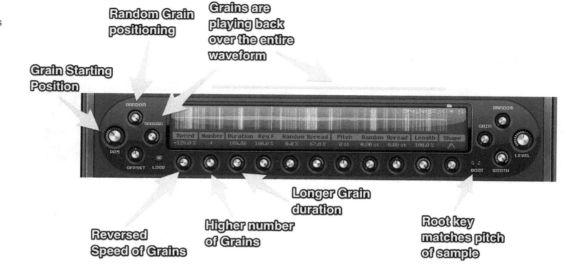

Figure 9.34 Grain parameters for Layer B in Padshop.

adds a distinctive pitch and a clear sustained character that was not present in Layer A. Because of its very different nature and function we set up some of the grain parameters in a very different way (Figure 9.34) compared with those of the previous layer.

As you can see, the position of the grain is set farther into the waveform in order to avoid the slow attack of the arco bass sample. We also used the Random parameter for the position of the grain in order to have the grains "jumping" at random locations inside the waveform. Through the Spread parameter we are allowing the grains to play back different parts of the waveforms. Notice how we are also using four grains (instead of one for Layer A) to add more excitement and richness to the sound. The Speed setting has a negative value so that the grains play back in reverse. Because this is a pitched sample we need to make sure that that Root Key is set to its pitch and octave. In this case we set it to G2. Let's listen to Layer B in Audio Example 9.40.

Envelopes

Now that we have set up the generators and the grains parameters, we can go back to more "familiar" territories with parameters and options with which, by now, we should know very well, such as filters and amplifier envelope. For a patch that is rhythmic in nature we need to make sure that the attack of the amplifier is short (Figure 9.35).

Notice how we set up the filters in different ways for the two generators: a HP for Layer A and a LP with a slow attack for Layer B. The slow attack on the LPF allows us to give a sense of motion and direction to the patch. To add more byte to the patch we added a tube distortion to both layers. Listen to Audio Example 9.41 to hear the patch with both layers activated and the filters and amplifier settings as shown in Figure 9.35.

Modulation

As was the case for the other types of synthesis we learned in this book, a sophisticated matrix of modulation can go a long way in creating more complex and interesting patches. With granular synthesis it is particularly important to have access to a comprehensive set

Figure 9.35
Filters and amplifier settings for Layers A and B in Padshop.

Figure 9.36
Modulation matrix for Layer B in Padshop.

of modulation sources and destinations. In addition to the usual destinations that we used to deal with from other types of synthesis (such as pitch, filter parameters, pan, etc.) in granular synthesis we can modulate some of the main grains' parameters such as duration, length, and speed. Assigning these parameters to different sources (such as modulation wheel, LFOs, or envelopes) allows us to morph the sound in real time and therefore make it richer and more interesting. For the rhythmic pad patch we are working on, we assigned the following modulation destinations: grains' duration, length, and speed to LFO, Key Follow, and Filter Envelope, respectively (Figure 9.36).

Listen to Audio Example 9.42 to hear Layer B with the modulation shown in Figure 9.36. As you can hear, the modulation adds a higher level of sonic complexity to the sound generated; a much richer tone overall.

For Layer A (the more rhythmic one of the two generators) we decided to use the Modulation wheel to control in real time the grain duration parameter. This allows us to morph in real time the layer from a controlled grain playback to a more chaotic one. Listen to Audio Example 9.43 to hear how the Modulation wheel changes the sound.

Using a step sequencer with a granular synthesizer can be very effective. It will give a bit more structure to the patch. If the patch you are designing sounds a bit too chaotic, modulating some parameters via a step sequencer can help. You can create more or less complex modulations that can affect the overall pitch or more advanced parameters such as the filter or the grain parameters. Take a look at Figure 9.37 to see how a simple pattern in alternating octave with a resolution of eighth notes can be used to modulate the pitch and the grain pitch of the two generators in Padshop.

For this patch we assigned the step sequencer to the pitch of Layer B and to the grain pitch of Layer A. Listen to Audio Example 9.44 to hear the sound with the step modulation applied.

Effects

Our patch would not be complete without the addition of some audio effects. In general granular synthesis allows us to add any sort of effects, from modulation to reverb, from distortion to delay, and so forth. Distortion and saturation can be helpful to enhance and highlight the "grunginess" of granular synthesis typical textures. On the other hand, a nice smooth chorus or flanger can go a long way to smooth out some of granular synthesis' harshness and abrasive

Figure 9.37
Example of simple step modulation in octaves with a synced resolution of 1/8.

nature. For pads we recommend adding a nice reverb and, if appropriate, a bit of delay in order to add some space to the patch. For this rhythmic patch we added a bit of reverb on Layers A and B (a bit more on the latter to make the sustained part of the patch sound a bit bigger). We also added a bit of chorus on Layer B. At this point make sure to balance the levels of the two generators. To make the patch more rhythmic, move the mix bar more toward Layer A, whereas, to make it more sustained and smoother, move it toward Layer B. For a more cohesive and slightly punchier pad we like to add a vintage compressor on the stereo output of the granular synthesizer (Figure 9.38).

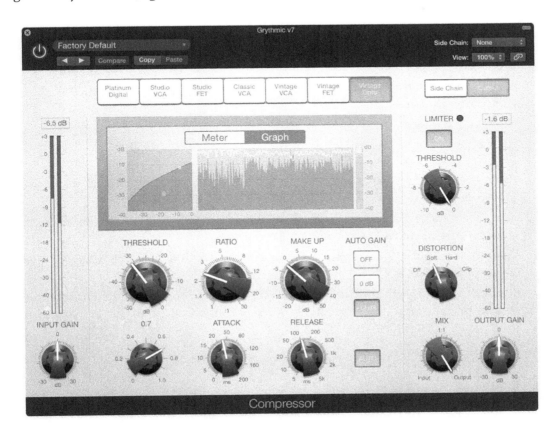

Figure 9.38
Vintage compressor in Logic's Pro X applied to the stereo output of Padshop.

Figure 9.39
Variation of the rhythmic pad.

To hear the final patch, listen to Audio Example 9.45.

Variations

One of the advantages of granular synthesis is that we can easily create variations from an existing patch that can sound mildly to drastically different from the starting one. Starting from our rhythmic pad that we created earlier, we can easily create a different patch by substituting Layer A (the acoustic guitar struck with a drumstick) with a more traditional drum loop, and on Layer B we can add a bit of pitch spread among the grains and increase the speed of the step sequencer. The final effect is a more rhythmically active and "pushing" pad, as you can hear in Audio Example 9.46.

Substituting the drum loop with a more sustained waveform allows us now to shift the pad more toward sci-fi sonority, yet maintaining a smooth rhythmic character. Specifically, we increased both the number of grains and the grain duration for Layer A (Figure 9.39) and added a delay effect, as you can hear in Audio Example 9.47. Notice how the waveform in Layer A is more sustained.

As you can see, granular synthesis is extremely versatile, powerful, and fun to program. It is virtually limitless when it comes to creating fun, fresh, and contemporary sonorities. Keep experimenting with it and don't be afraid of trying new solutions and even some extreme ideas: granular synthesis will always give you something back worth using in your productions!

Exercises

9.1 Using a wavetable synthesizer, create two patches of your choice from the list shown in Table 9.4.

9.2 Using a wavetable synthesizer, create a patch that is totally "outside the box"; don't be afraid to experiment with unconventional waveforms.

9.3 Using your granular synthesizer of choice, create a rhythmic pad following the steps shown in this chapter.

9.4 Starting from the patch created in the previous exercise create three different variations that are related, but sound substantially different from the original.

INDEX

Ableton, 155, 157, 159
Air Music Technology, Loom, 184, 186, 187
Aftertouch, 96–97, 99–100, 154, 185, 205, 236
Akai
    EWI, 109, 264–265, 269
    MPC60, 25, 216–217, 247
    S-900, 216
Algorithms (FM), 143–145, 150–151, 153, 155, 157, 168, 209, 246
Allpass Filter, 66, 83
Alonso, Sydney, 21, 26
American Composers' Alliance, 5, 13
Analog Modeling, 263
Anderson, Laurie, 25
Appleton, Jon, 21, 26
Applied Acoustics Systems, 116
    String Studio, 269, 275, 277–280, 284–287, 289
    Tassman, 28, 73–74, 116, 118–119, 127, 155, 177, 264–266, 301
ARP
    ARP 2600, 70, 75, 125, 127
    Odyssey, 16, 28, 125
Arpeggiator, 88, 104, 165–166, 285, 287, 309–312
Arturia V Collection, 127
    ARP 2600, 70, 75, 125
    Jupiter-8V, 2–3, 70, 86–88, 137
    Minimoog V, 41, 45, 54, 56, 59, 86–87
    Oberheim SEM, 62, 70, 89
    Prophet V, 60, 67, 119, 120, 295,
Audio Engineering Society (AES), 19, 22, 29

Bandpass Filter, 64
Banks, Tony, 13
Beatbox Kit, 246–261
Beating, 41, 44–45
The Beatles, 3–4, 15, 84, 204
Belar, Herbert, 6, 10
Bralower, Jimmy, 25
Breath Controller, 100, 109, 138, 264
Buchla Electronic Musical Instruments, 4, 13, 14, 88, 115,
Buchla, Don, 11, 13, 14

Carlos, Wendy, 13, 182
Carrier Frequency (FM), 23, 141–151, 155, 158, 160, 164, 168–169
Casio FX-1, 216
Chamberlin, Harry, 15

Chorus Effect, 43, 83–86, 162, 185, 192, 196, 198, 309, 313
Chowning, John, 22, 28
Clapton, Eric, 25
Columbia-Princeton Electronic Music Center, 4, 6, 10–11
Compression (sound pressure), 31–32, 55
Compression (dynamics effects processing), 111, 159, 194, 196, 198, 211, 214, 223, 238, 250, 252
Control Changes (CC), 96–99, 108
Control Voltage (CV), 75–76
Controller (MIDI), 100–103, 105, 107–110, 113, 116, 122, 124, 134, 138, 151
Convolution Reverb, 81–82, 200–201
Cornell University, 12
Critical Bands, 44
CrusherX-Live, 315–316
The Cure, 183
Cutoff (frequency), 61–64, 68, 73, 82, 87, 89, 101

Detuning (oscillators), 38, 44, 85, 119, 136, 145, 150
Deutsch, Herbert, 12
Digital Control Bus (DCB, Roland), 20
Digitally Controlled Oscillator (DCO), 21, 23
Distortion (effect), 46, 76–77
Dolby, Thomas, 19, 21, 23–24, 292
Doppler, 41–42
Duophonic/Duophony, 16, 28

E-MU, 24
    Emulator I, 24, 28, 216
    Emulator II, 216
eSATA, 93
Echo (effect), 77–78
Electric Piano (FM), 143, 152, 153, 154, 159
Emerick, Geoff, 3
Emerson, Keith, 13, 125
Ensoniq Mirage, 216
Envelope Generator (EG), 12, 17, 22, 56, 71, 175,
Expression, 99, 100, 109

Fairlight, 21, 23–24, 28
Fast Fourier Transform (FFT), 176
Firewire (IEEE 1394), 21, 105, 106
Flanger (effect), 135, 168, 185, 192, 196, 197
Foot Controller, 100, 205
Fourier Synthesis, 177
Fundamental (frequency), 32–38

Index

Gabriel, Peter, 24, 25
Gate (CV), 74, 76, 87
Genesis, 13, 183
Glissando, 86
Guitar-to-MIDI, 108

Hall & Oates, 25
Hammond, Laurens, 8, 18
Hammond Organ, 8
Harmonic Series, 34–38, 50–57
Henry, Pierre, 4

iAcoustica Drum Library, 217–224
iMPC, 247–252
Image Line, Morphine (instrument), 184, 186, 187
Impulse Response (IR), 82–83, 201
Infrasonic, 141–142
Inharmonic, 34–35, 51, 143, 175, 278,
Izotope IRIS, 69, 70, 71, 181

Kakehashi, Ikutaro, 20
Kawai, 95
    K5, 28, 182, 183
    K5000, 178, 182, 183
Korg, 16, 95, 294
    Kronos, 154
    Legacy Analog Collection, 127
    M1, 27–28
    Polysix, 126
    Wavestation, 27, 296, 300
Kurzweil
    K150, 28, 177, 182, 183, 184
    Kurzweil 250, 216

Lauper, Cyndi, 25
Linear Arithmetic Synthesis, 294
Linn LM-1 Drum Computer ("Linn Drum"), 24–25, 28
Logic Pro X, 28
    Alchemy, 178–181, 184, 186–189, 192, 195
    EFM1, 145, 155
    ES1, 59, 61–63, 127
    ES2, 26, 64, 69, 87, 121–122, 127, 292–293
    EXS24, 204–210
    RetroSynth, 127
    Sculpture, 267–275, 281–283
    Space Designer, 201
Lookup Table, 292–293, 296
Loudness, 46–50
Low-Frequency Oscillator (LFO), 68–70, 71, 73
Low-Pass Filter (LPF), 2, 61–64, 69

Madonna, 25, 152, 155, 160
Martin, George, 3–5
Mellotron, 4, 14–15, 19, 23, 204–205

Masking (frequency), 45–46, 48, 82
Modulation Frequency, 144–147
Modulation Index, 149
Modulation Matrix, 87, 308, 310–312, 322
Modulation Wheel, 89
Modulator (FM), 141–145
Moog Music, 7, 8, 14
    Little Phatty, 126,
    Minimoog, 4, 19, 124, 125
    Modular, 2, 12, 13, 68, 115
    Polymoog, 17, 28
    Prodigy, 125
    Sub 37, 16
    Voyager, 85–86, 126
Moog, Robert, 11–13, 14
Muse Research Receptor, 94–95
Musique Concrète, 3–4, 6

NAMM (National Association of Music Merchants), 20
Nave, 292, 301
Native Instruments, 42, 155
    Absynth, 67
    FM8, 145–148, 161–168
    Kontakt, 204–205, 208–209, 225–261
    Massive, 59, 67, 70, 292–294, 297
    Reaktor, 28, 73, 177, 301, 314
New England Digital (see Synclavier), 21, 23, 26, 182, 204
Nintendo Wii, 13
Noise Gate, 197
Nord, Lead, 28, 126, 154, 263
Notch Filter, 65
Note On/Off, 19, 95–99, 108

Oberheim, Parallel Bus, 20
Oberheim, Tom, 63
Olson, Harry, 10
Ondes Martenot, 4, 7
Operators (FM), 144–145
Oscillator (see Voltage-Controlled Oscillator), 53–59
Oscilloscope, 188–189

Padshop, 314, 316–323
Paraphony/Paraphonic, 17–18
Passband, 62–64
Pattern Sequencer, 88–89
Phantom Image, 43–44
Phase, 36–38, 41, 62, 66
Phaser, 83–85, 192
Pink Floyd, 7, 13
Pink Noise, 54, 58–59, 64
Pitch Bend, 89–90, 96–99
Polyphony, 10, 15–19
Portamento, 86–87, 100–102